HARDY APPLES

HARDY APPLES

Growing Apples in Cold Climates

Bob Osborne

Photographs by Beth Powning

FIREFLY BOOKS

Published by Firefly Books Ltd. 2022

First printing

Library of Congress Control Number: 2021951195

Library and Archives Canada Cataloguing in Publication
A CIP record for this title is available from Library and Archives Canada

Published in the United States by
Firefly Books (U.S.) Inc.
P.O. Box 1338, Ellicott Station
Buffalo, New York 14205

Published in Canada by
Firefly Books Ltd.
50 Staples Avenue, Unit 1
Richmond Hill, Ontario L4B 0A7

Cover and interior design: Hartley Millson
Illustrations: George A. Walker
Editor: Julie Takasaki
Copyeditor: Ronnie Shuker
Proofreader: Jennifer D. Foster
Indexer: Siusan Moffatt

Printed in China

We acknowledge the financial support of the Government of Canada.

Dedication

This book is dedicated to my patient wife, who has seen less of me than my orchard has, and to the community of apple explorers who have helped to save countless cultivars from oblivion so future generations can benefit from the vast richness of the apple world.

Acknowledgments

I would like to thank the following people for their knowledge, patience, help and enthusiasm:
Firefly Books, for their belief in the project.
Julie Takasaki, my supportive and exacting editor.
Ronnie Shuker, my copyeditor, for his detailed editing and encouraging words.
Garth Nickerson and Chris Maund, for their comments and insights on the text.
Blair Stirling, for his help with insect control.
Belliveau Orchards, for their hospitality and knowledge.
Vern Grubinger, for permission to recreate the table in Chapter 4.
John Bunker, Veronica Carl of Copenhaven Farms, Tianna DuPont, Claude Jolicoeur and Prairie Hardy Nursery, for their photos.
Beth Powning, for her artistry and dedication.

Additional photo credits

Alamy Stock Photo
Babette French: 191; Cécile P./Stockimo: 242; Felix Choo: 246; Graham Corney: 271; MPB-one: 269 (right); Nigel Cattlin: 91 (top); Tim Gainey: 48.

Bob Osborne: 94 (top), 109, 119, 168, 247, 256, 272.

Claude Jolicoeur: 268, 269 (left), 271 (right), 274 (right), 275.

David L. Hansen, University of Minnesota: 178, 198, 243, 264, 265.

iStock Photo
mschowe: 211; Veronika Roosima: 274 (left).

John Bunker: 148, 153, 154, 189, 273.

Prairie Hardy Nursery: 160, 217.

Shutterstock
aleori: 91 (bottom), 93; Andrei Metelev: 40; Dusan Petkovic: 101; Erik Agar: 73 (bottom); Greenseas: 32; iPostnikov: 10; Marija Stepanovic: 80 (top); Tomasz Klejdysz: 87; Werner Rebel: 64.

Tianna DuPont, WSU Extension: 90.

USDA Pomological Watercolor Collection, Special Collections, USDA National Agricultural Library: 14.

Veronica Carl: 112, 115.

Wikimedia Commons
Eugene Zelenko (CC-BY-SA-4.0): 75 (top); PVM (CC-ASA-4.0): 164; Wikimedia Commons (PD-US-expired): 13.

Contents

Introduction

I grew up in a home surrounded by fruit trees, roses, shrubs and perennials. My thinking places were under thorny quince bushes and at the top of the maple tree. Fruits of all kinds grew in the yard. My neighbor hosted the morning gardening show on the radio, and his yard was a maze of grapes, apples and raspberries. Though I began my career in a different field as a cabinetmaker, I was drawn inexorably toward growing.

My life changed when I first learned how to graft. It was as if a spell had been cast about me, and from that point on I was determined to grow fruit trees. It was the apple I was most drawn to.

This book is a distillation of what I have learned after propagating and growing apples for 40 years. Having moved to an area where temperatures can drop to –40°F (–40°C), I quickly learned there were many apples that would not put up with such frigid winters. My passion became to learn as much as I could about the cultivars and rootstocks that would.

This is not as much a growing guide as it is an homage to the apples that exist in such wonderful abundance and diversity. Many people will never get to savor the sweet juice of a Fameuse or the nutty complexities of a Frostbite, and that is a pity. This book is an effort to make as many readers as possible aware that

◂ Trees in full bloom in the seed orchard at Corn Hill Nursery.

these wonderful apples exist and to compel you to find and grow them. However, even those who will never grow them can find pleasure in hearing their stories. In addition, it is my hope that the descriptions and photos of the fruit, leaves and wood can be used to identify apples, particularly those growing in older orchards. I would have loved to have had such a guide when I was first learning about apples.

We have never used pesticides, herbicides or synthetic fertilizers in the care of our apple orchard. Our orchard's purpose was to provide us with wood for grafting, so the quality of the fruit has never been an issue. We are not apple growers; we are apple tree propagators. Yet because we have not interfered with sprays in the orchard for 40 years, we have had the unique experience of seeing how different apples grow under modified neglect. Some are plagued with fungal diseases such as apple scab every year, while others show varying degrees of resistance. Some drop their fruit early. Some hang on till the dead of winter when my son and I pluck them on warmer days to suck out the slightly fermented juice. There are apples that can be harvested for our table that look as good as any fruit in the supermarket and some only fit for cider.

Such experience has given me a tremendous respect for apple growers. Even with every available chemical tool, growing quality fruit is not an easy process. Growing fruit organically is perhaps the most challenging task in horticulture.

The past century the business of apple growing has become dependent on a host of products, many of which are harmful to the environment and the people who use them. Today, we are witnessing a concerted effort to develop new, more life-affirming methods of production. This book describes some of the newer products available to growers, and both old and new methods that can help decrease the ravages of apple pests.

The home gardener can gather whatever they deem suitable in this book for their small orchard. Hopefully it will help you grow fruit you can be proud of, but even blemished fruit can be used for cider or cooking. With a bit of effort and time you can produce enough good-quality fruit for your family's needs and beyond.

No matter how many trees you grow or what level of maintenance you adopt, growing fruit is always rewarding and enlightening. Continuing the tradition of growing food for yourself and others strengthens our communities and the security of our food supply. Today, when few people know where their food originates, it is more important than ever to regain that knowledge and rekindle a passion for gardening and farming. There is nothing more important than a source of healthy food.

Bob Osborne
Corn Hill, New Brunswick

Cultivar vs. Variety

The seeds of an apple will not produce trees identical, or often even similar, to the mother tree. There are approximately 57,000 genes in the apple genome, twice the number in humans. This means that when one apple is crossed with another, the resulting seedling has a mind-boggling number of combinations and permutations that will decide its characteristics. If you want to keep the exact characteristics of an apple you cannot do this by growing its seed. It must be propagated by grafting a piece of its wood onto a seedling or, rarely, by growing it from a cutting. In this way the McIntosh apple grown today has remained, more or less, genetically identical to the seedling found in 1796 in Dundela, Ontario.

When talking about a specific apple, most people refer to the "variety." In botanical terms this is incorrect. A variety refers to a group of seedlings within a species that appear in nearly all respects the same. A good example is the vegetables we grow from seed. A Green Arrow pea will look like every other Green Arrow pea, but genetically speaking they are not identical. A "cultivar" refers to a seedling that, because of its desirable characteristics, is named and then it is propagated by asexual means to maintain its unique set of genes.

Though it is a departure from the language of most apple books, I will refer to what most people call a variety as a cultivar.

The Golden Delicious apples that grow on my trees will look different than Golden Delicious from other areas.

Information and Descriptions

Though I have spent half a lifetime working with apples, I feel I still know very little. The old adage "the more you know, the less you know" is made apparent every day. Much of the information in these pages has been gleaned from experience, but no person can know more than a portion of any subject, so I have had to reach out to many sources of information to fill in the gaps. Though I have tried to ensure that the information is accurate, I often rely on the knowledge of others. There are bound to be

inaccuracies, and I apologize for any mistakes and request that any corrections be relayed to me. Learning should never end.

It should be noted that apples can behave very differently in different places. Geography, topography, climatic patterns, soils and management practices can cause an apple to prosper in one place and suffer in another — sometimes within a dog's bark of each other. An apple grown in a warmer climate under irrigation and intensive management may look and taste much different than the same apple grown where it is cold and under less-intensive management. My Golden Delicious barely resembles those from Washington State — same genetics, different results. That said, the descriptions in this book are based on my experience with these particular cultivars, and you might have a vastly different experience with those same apples. I consider this part of the joy of cultivating fruit trees — every season, every fruit invites a new and unique moment.

1 Origins

The story of the cultivated apple begins where the vast steppes of Central Asia buckle against the steep rocky slopes of the Tian Shan Mountains. Even today remnants of apple forests can be found in the foothills of the Tian Shan in what is today's southern Kazakhstan. The apples grew from the lowlands, where summers are hot and dry and winters cool, to the high valleys of the mountains, where summers are cool and winters bitterly cold. This diversity of habitats has created a diversity of genetics, and this is reflected in the immense geographic range of the apple. Apples are grown in places as different as North Africa and Siberia.

The original home of the apple was once an important stopover for the caravans of Bactrian camels and horses burdened with goods being transported between India, Sogdia, Persia, Europe and the vast empire of China. It was here that the Saka and Wusun cultures lived. They drove immense herds of cattle and horses pastured on the grasslands of the steppes. While most maintained their herds, some farmed the alluvial fans, which were formed by braided rivers carrying meltwater from the perpetually white peaks that rose above them. And there would have been apple

◂ Apple trees blossom near Almaty, Kazakhstan.

pickers. Here caravans took on dried apples, apple leather, hard cider, apple seeds and perhaps some fresh apples for the trip, if it was summer or fall. If heading toward China, the caravans would pass through inhospitable lands, including deserts that seared flesh in the day and froze it at night. No apple could remain fresh for long in such a place.

This ancient web of roads that we now call the Silk Road was where goods such as silk, jewels, ivory and other precious items traveled between civilizations, but what was perhaps most important was the spread of cultural information, which included the foods and agricultural techniques of other peoples.

Rhubarb made its way from China and grapes from Persia. The apple found its way both east and west. It traveled eastward across the desolate lands of the Dzungarian Basin and the Taklamakan Desert, passing through a chain of oases till it reached China, where it has been an important fruit for centuries and where the art of grafting is said to have originated (most likely grafting started with peaches, which are native there). The apple also made its way westward along the trade routes to Persia, then northwest across Asia Minor to Europe, where it thrived and was soon embraced as an important food. From Europe it traveled by boat to the colonies of what we now call North America. Although species of native apples exist in North America, all are exceedingly bitter and extremely small. Essentially, all edible apples grown in North America, and throughout the world, are descendants of apples that once grew on the flanks of the Tian Shan, though recent DNA evidence has shown that some genetics have come from *Malus sylvestris* (European Crabapple), a species from Eastern Europe and environs. There would have been pollen being transferred back and forth between the local species and the lately arrived *Malus sieversii* of the Tian Shan. Seedlings would then have shared genetics.

To the new settlers of North America, a fruit tree was a symbol of home. Initially, apples brought from England, France and Holland to the Americas were transported as seeds gathered from favorite trees in the homeland. An ocean voyage could last weeks or months, and grafted apples were difficult to keep alive in such conditions, so the spread of apples on the continent was initially slow.

Those first seeds had many factors working against their success. First the person carrying the seed had to plant it. Even when this happened, it was essential that certain conditions were satisfied. Without a cool moist period called stratification, an apple seed may not break out of its dormant state and will eventually die. It is likely that most seeds never germinated. Those that did live had to survive trials such as animal browsing, weed competition, drought, poor soil and neglect. The weather patterns and harsh winters posed further hurdles. The winters of northern areas or the humidity and heat of the southern areas acted to select only those seedlings that could adapt to the environments they were planted in. Only after such trees started to produce their own seeds could the proliferation of the apple begin on any scale.

When colonization increased and transatlantic voyages became more numerous, apple cultivars from Europe gradually began to be transported as young grafted trees. In this way the apples of France found their way to Quebec. The apples of England arrived in the colonies that would become the United States. As well, they found their way to Atlantic Canada and Ontario (then called Upper Canada). Many European apples failed in their new homes, but others prospered and were added to the growing assemblage of "New World" apples.

The movement of apples across the North American continent took place in myriad ways. One man became a legendary figure as a result of his exploits. John Chapman (1774–1845), a traveler and proselytizer of the Swedenborgian sect of Christianity, roamed the areas of Ohio and Indiana. He journeyed through the countryside spreading his gospel in bare feet, even in winter, and dressed in coarse clothing, believing that suffering in this life would bring comfort in the hereafter.

He became the first large-scale nurseryman in America and was known as Johnny Appleseed. He traveled down the Ohio River system, finding suitable alluvial land along the river that could be worked easily. Once he started the nurseries, which often contained hundreds of thousands of seedlings, he left them in the care of others, who sold the apples to homesteaders for approximately 6 cents each.

Chapman capitalized on the settlers desire to plant apple trees in the lands they were traveling toward. He became symbolic of the journey of apples throughout the continent. He was certainly one of the most important and prolific disseminators of apples in the expanding territory of what is now known as the United States of America. Interestingly, he had religious objections to grafting, and the immense nurseries he planted were grown entirely from seed, as were most of the first apples in the newly cleared lands.

An etching of Johnny Appleseed.

Many apples produced by seed were too sour, bitter, dry or coarse to qualify as dessert or even cooking apples. Most were used to make cider. This alcoholic drink was as good as currency for pioneers. Alcohol was in constant demand. Hard cider was not as perishable as the fruit. This was long before refrigeration,

Watercolors of Alexander (left) and Rhode Island Greening (right) from the U.S. Department of Agriculture Pomological Watercolor Collection, a botanical resource that documented fruit and nut varieties between 1886 and 1942.

when the only means of keeping apples was to dry them, store them in cool root cellars and basements or ferment them into hard cider.

Apples soon became an important food crop in both the United States and Canada, and by the late 19th century hundreds, if not thousands, of named cultivars were being grown by nurserymen and farmers across the continent. In the more northern areas, however, the harsh winter conditions killed many of these trees.

A series of imports had a huge impact on the apple mix for the north. In the 1880s apples that originated in the colder areas of Russia were brought to Canada and the northern states. Names such as Yellow Transparent, Tetovsky, Antonovka and Alexander quickly became household names as a result of their hardiness and tenacity.

It was during this period that both private individuals and public institutions started breeding apples in earnest, crossing hardier types with less hardy but higher-quality cultivars. As well, many new apples were the result of people discovering superior seedlings in their hedgerows and orchards. The expanding stable of apples resulted in better-quality fruit and apples that could be grown in areas where very low temperatures had earlier stymied fruit growing attempts. Cultivars such as Wealthy, McIntosh, Dudley, Red Astrachan, Fameuse, Northwestern Greening, Bethel and New

Brunswicker were soon the apples of choice for cold country orchards.

The process of breeding continues today, most often in institutional settings. Apples bred for quality and hardiness appear in ever-increasing numbers. We see cultivars such as Honeycrisp, Honeygold and Keepsake from the University of Minnesota; Spartan, Sandow and Sunrise from Agriculture Canada; and Cortland, Empire, Liberty and Macoun from Cornell University.

Some of the most exciting new introductions are the result of collaborations between several institutions that are breeding apples with resistance to disease, most importantly apple scab (caused by the fungus *Venturia inaequalis*). The beginning of this story goes back to the observations of L.F. Hough, a graduate student at the University of Illinois in 1948. Hough had been investigating the research done by Dr. C.S. Crandall, who had been crossing crabapple species with commercial apples. A severe infestation of scab had nearly defoliated the doctor's orchard that year. One tree caught Hough's attention, a seedling that had resulted from crossing *Malus floribunda*, a crabapple species, with the cultivar Rome. It was untouched by the disease. Hough wrote a paper hypothesizing that this seedling (*Malus floribunda* 821 x Rome) could be used to confer resistance to seedlings.

In concert with Dr. J.R. Shay of Purdue University, Hough began a breeding program using the scab-resistant seedling. Because of this effort, we now have apples with superb quality and resistance to a multitude of fungal diseases. The program was named PRI, which stands for Purdue University, Rutgers University and the University of Illinois — the first institutions to begin the work. Soon many universities and fruit breeding stations joined in the process of creating new cultivars that were disease-resistant, and the work continues.

And we should never discount the curious people who taste wild apples and realize their potential. The world of apples is filled with chance encounters that became household names when their qualities were appreciated by an ever-increasing number of apple eaters.

Today breeding work continues to enhance quality, hardiness, storability, and disease and insect resistance. An interesting line of work involves using materials gathered from the apple's homeland, Kazakhstan. It was a Russian scientist, Nikolai Vavilov, who first alerted the world that he had found the original home of the apple. Though he was initially ignored, eventually others came and confirmed what was obvious.

There is an immense diversity of apples in the Kazakh forests. Apples of every size, shape, color and description still grow on the slopes of the Tian Shan. Researchers from Cornell University, among others, have gathered scions (detached buds or shoots used in grafting) and seeds from hundreds of trees and are working with this material to create an entirely new gene pool and, thus, an entirely new generation of apples. The apple has come full circle.

2 The Apple Tree

Every apple tree starts from a seed. A seedling can grow to be a towering giant, a diminutive shrub or, most likely, something in between. The average apple tree grows between 16 and 32 ft. (5 and 10 m) high. If open grown, once the tree has cropped, it will usually end up a rounded, spreading, slightly pendulous tree with a stout strong trunk, or even several trunks. Apple trees come in all sizes, but once fixed in your mind, it is easy to pick out their silhouettes in the landscape.

Rugged apple trees, planted long ago when butter came from churns and horses plowed the fields, can still be found in unsung places, lost among the hills and surrounded by the seedlings they have thrown. Some are cared for yet by farmers, descendants of the women and men who churned and plowed. They may still pick apples from trees that helped their families live through many a lean winter. And then there are the wild apples, whose seeds were spread by birds or by people tossing the remains of a snack into a pasture or along the edge of a hedgerow or forest, where there is light and a chance to prosper.

◂ Roots, shoots and leaves all serve to create the fruit. For the tree, the fruit is the seed for procreation, for us a tasty treat.

A typical low-density orchard in spring.

No matter where you find apples, there is the anticipation of the ripening fruit, and when the time is right — the picking. Bins of yellow, red and green spheres resting in baskets and pails and cardboard boxes. The smell as they are handed up onto a truck or wagon. The feel of the first crackling bite of the year that leaves the taste of juice, tart and sweet, upon the tongue. To pick the fruit of trees you have grown is a gift from life both precious and sublime. For those drawn to grow the apple, few of life's choice moments can compare to the quiet satisfaction gained by seeing through the growth of a sapling into a tree that brims with fruit.

This chapter will explore the various parts of the apple tree, from root to fruit, and explain how they function to produce the trees we admire and the apples we savor.

The Roots

Because we walk on the surface of the soil, we tend to notice only the things we can see, touch, hear and smell. However, below the surface is a world just as vital as that above ground. The roots of an apple tree provide both anchorage and sustenance to the tree.

Often, there is nearly as much of the tree underground as above.

To provide anchorage and sustenance, apple tree roots must penetrate the spaces between the particles of soil. At the tip of every root is a group of cells called the root meristem. Generally speaking, meristem cells are undifferentiated cells that are capable of cell division and are responsible for the continuous growth of the plant. The root meristem's function is to create new cells that will enable the root to push deeper into the soil. At the very tip of the root is the root cap, which acts like the point of a needle, pushing apart the soil particles while at the same time protecting the meristem cells from injury. As the cap pushes through the soil, many cells are sloughed off through friction, but new cells constantly replace them. The actual lengthening of the root is the result of the elongation of cells created by the root meristem, and it is the growth of these cells that pushes the root forward. This section of the root is called the region of elongation. Behind this region, closer to the tree's trunk, the outermost cells of the root (also known as the epidermal layer) form thread-like extensions called root hairs. Their main function is to absorb water and nutrients from the soil.

There are several processes involved in the absorption of water and nutrients. Roots can absorb these directly, but they also form connections with a group of soil fungi called mycorrhizae. The mycorrhizae deliver water and nutrients to the roots in exchange for sugars. This symbiotic (mutually beneficial) relationship allows the tree to access water and nutrients from a far larger area than it would without such assistance.

The most important source of nutrients is the soil water, in which many elements are held in solution. These nutrients are the result of the breakdown of organic material by bacteria and other soil organisms. The resultant "manure" and the decomposition of dead soil organisms are keys to optimal soil health.

As with any structure, a strong foundation is important for long-term survival. If your soil is open and friable (easily crumbled) the roots will spend less energy trying to push through it; if your soil is deep the tree will be able to form strong anchoring roots; and if your soil is rich in nutrients the roots will be able to provide the necessary elements to create a healthy tree.

The Trunk and Branches

It is reasonable to think of a tree as a solid living thing, yet a tree is more correctly an ever-expanding skeleton of dead wood encased within a thin layer of actual living and producing cells — a layer perhaps the thickness of cardboard. Nearly all the rest is composed of dead cells that now act only as support.

Beneath the bark lies a thin layer called the cambium. This layer is the only living part of the tree and any damage done to it will affect the tree's ability to provide water and nutrients to the canopy and the roots.

Within the cambium, cells divide and multiply toward the outside of the tree to form phloem cells, which transport sugars throughout the tree. These cells are short lived, and

Each ring consists of the lighter spring wood and the denser, darker summer wood. In this small branch you can see the initiation of the dark heartwood.

when they die they form a new layer of bark. Toward the inside of the tree, cells are created called xylem cells. These are tube-like cells with sieve ends that allow water and accompanying nutrients to rise from the roots to the canopy. When xylem cells die they harden to form what we refer to as wood, which provides both strength and flexibility to the tree.

As the trunk grows upward, lateral branches are initiated. The branch-growing process is governed by meristem cells at the terminal bud of the central leader (main upright stem). The shoot meristem cells produce hormones called auxins that elongate the cells of the terminal bud upward but also suppress the initiation of new lateral branches until the terminal bud has grown far enough away. Once the auxin concentration drops to a level that no longer suppresses growth, a new branch is initiated. The exact spacing of these new branches depends on the genetic makeup of the individual cultivar and the rapidity of growth. This internal self-regulating system prevents the tree from producing too many branches, which would diminish light levels in the canopy and therefore lessen food production. Isn't nature smart?

If you split a trunk longitudinally, you can see the initiation point at the very center of the tree. Once growth of a new branch begins, any new branches forming off that branch will initiate at the meristem in the same manner.

It should be noted that the lateral branches of a tree do not grow upward as the tree ages but stay in their same position, only growing in girth. This is an important factor to understand when deciding on how to prune a tree.

The direction of the branch is influenced by light availability. Growth will always move toward high light levels. This is why it is important to keep trees open to light on all sides so that the growth will be even in all directions, which creates the highest canopy volume.

If a trunk or branch is cut across its length, the growth can be seen as rings. Each ring is the result of a year's growth. The yearly layer is itself composed of the more open spring wood and the tighter summer wood, giving each ring a two-toned appearance. The apple's young wood is ivory white. The older heartwood is usually a dark reddish brown, a rich color prized by certain makers of furniture, bowls and such, yet rarely used commercially because apple logs are seldom of any length and are usually twisted.

Though similar in form, apple leaves can vary in size depending on the cultivar, and the edge serrations can be sharp or more rounded.

The Leaves

Apple leaves are arranged alternately on the stems. They are ovate in shape and end in a pointed tip. The edges are serrated — usually with pointed serrations, but they can be more rounded. There is a main central midrib vein with alternating smaller veins running to the leaf edge. The attachment of the leaf to the stem is called the petiole. Its length and thickness can vary, but most petioles are slender and around 0.8 to 1.2 in. (2 to 3 cm) in length. It links the leaf to the stem, both bringing water and nutrients to the leaf and delivering sugars produced by the leaf to the stem and, thereafter, the rest of the tree.

Though relatively simple looking, what goes on within the leaf is far from simple. The surface is composed of a single layer of epidermal cells. The upper layer is coated with a waxy covering called the cuticle, which helps prevent water loss and plays an important part in preventing disease entry. The undersides of the leaves are softer looking, with fine structures that give it a slightly woolly look. Scattered across this surface are stomata, which are tiny openings. These are controlled by two opposing cells that open and close depending on the humidity in the air and the amount of moisture in the leaf. The stomata also control the exchange of gases in and out of the leaf.

Under the upper epidermal layer is a layer of parenchyma cells. These cells house specialized structures called chloroplasts, which contain the magical substance chlorophyll. By absorbing and using the energy of sunlight, chlorophyll converts the low-energy elements of water and carbon dioxide into high-energy food — sugars — and releases oxygen as a by-product. Once sugars are created, they travel back through the petiole and are distributed to wherever growth is taking place in the tree.

It is important to remember, especially when you are pruning, that the roots are the last section of the tree to receive these sugars. When sugar production in the leaves does not leave enough energy for both top and root, it is the root that will suffer and, as a consequence, the top will not receive what it needs. This can lead to decline and will weaken the defenses of the tree. Pruning, as you will discover, is always about balance.

Apple leaves can be useful tools for identifying a cultivar. The shape of the leaf and the appearance of the edge serrations are like fingerprints. The photo on the left shows a few leaves to illustrate the differences that can be found in terms of size, shape and edge serration.

Most apple blossoms are light pink to white.
Nearly all are fragrant.

The Flowers and Fruit

The flower is the organ that has evolved to transfer pollen from one tree to the waiting receptacles of other trees of its kind so that fertilization and seed development can occur. The parts of a flower are specialized leaves that evolved millennia ago and no longer resemble leaves at all.

Apple flower buds form the summer or fall before they open. They can be located on a specialized branch called a spur, either singly or in clusters, or occasionally at the terminal ends of stems, depending on the cultivar. The number of flowers produced in a given year depends on the cultivar's habit — some bear annually, while others produce good crops every other year — but the number of flowers can also be

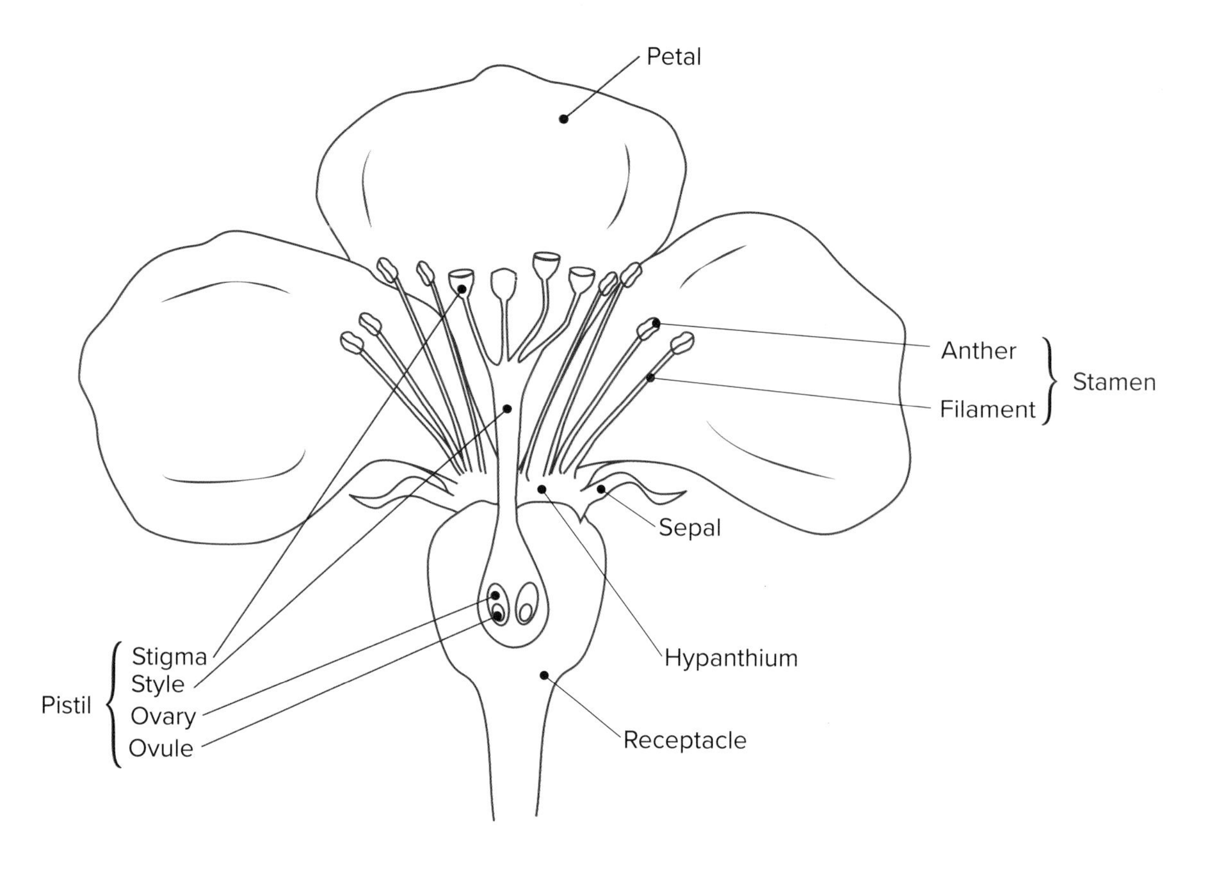

influenced by age or stresses, such as weather or soil conditions.

During winter bud scales protect the latent flower. As the bud expands in the spring, a flower emerges, supported by a stalk that widens into the receptacle. Above the receptacle are five sepals, which are specialized leaf structures, and the hypanthium, which is the flesh that unites all the parts of the flower. The apple flower has five petals. These are generally white to light pink, though some crabapple blossoms can have deep-pink or reddish petals.

Inside the petals of the flower are two different reproductive structures. Surrounding the center are several stamens. Each stamen has a slender filament that supports an anther, which produces pollen. In the very center of the flower is the pistil, composed of the stigma, style and ovary. There are five stigmas on the top of the pistil, and they produce a sticky substance, which traps the pollen that falls off pollinators or arrives by wind.

Once a pollen grain lands on a stigma it germinates and grows a pollen tube down through the style. Each pollen grain contains two sperm. When these arrive at the ovary, one sperm unites with the egg in the ovule and the second unites with two specialized cells (haploid cells) in the same ovule. The fertilized egg begins cell division and becomes known as a zygote, eventually forming an embryo. The second union produces an endosperm, which acts as food for the developing egg.

The growing embryo and endosperm are encased by the remaining portion of the ovule. This becomes the seed coat for the developing

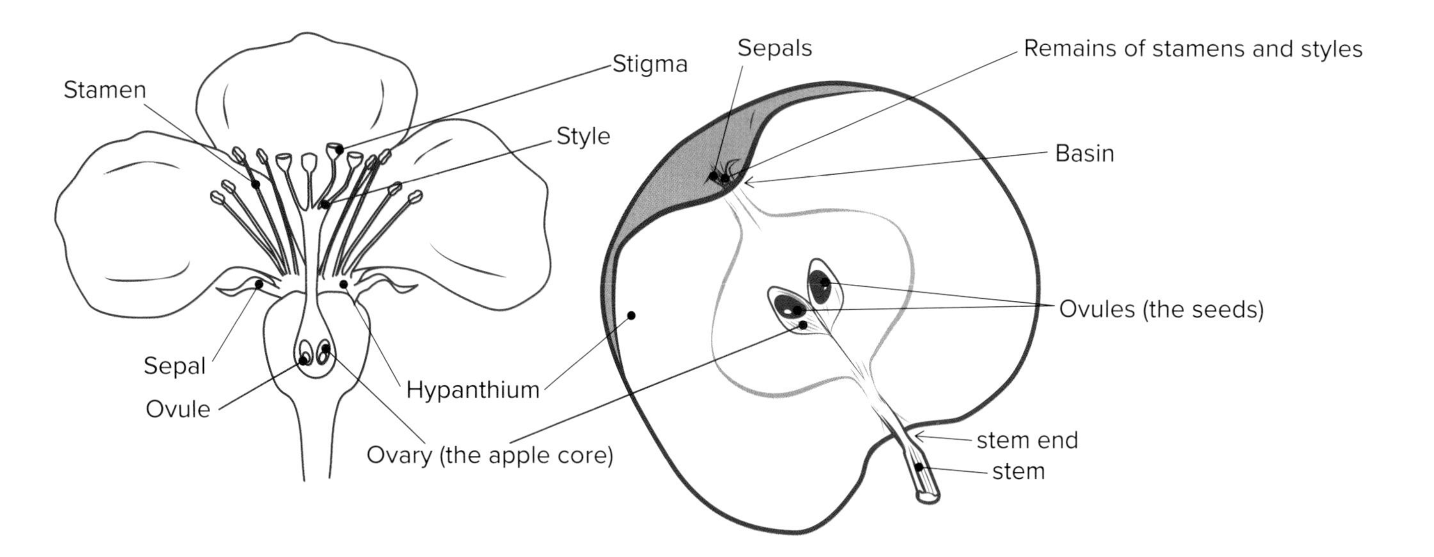

seed. The endosperm shrinks as it forms two cotyledons that fill most of the seed. The cotyledons will eventually feed the tiny embryo that is located at the narrow tip of the seed. The embryo forms a meristem that will divide its cells to form both a shoot and a root. Upon germination, the cotyledons feed the embryo until true leaves are formed that can begin producing food for the seedling.

The remains of the ovary form a core around the seeds whose limits are defined by a thin layer of tissue, which is usually somewhat darker than the flesh beyond it. An apple has five core sections and can produce up to 10 seeds (though most average five to eight per fruit). If fewer seeds are produced, the fruits will be proportionally smaller.

As the fruit develops, the hypanthium — the fleshy tissue that unites the sepals, stamens, pistil and petals — forms the pome surrounding the core, the portion we crave. The remaining sepals form the calyx, which can be seen on the end of the apple, opposite the stem cavity.

While apples are similar in many ways, their size, shape and color differ from cultivar to cultivar. Qualities, like the length and thickness of the stem and the depth and width of both the stem cavity and basin cavity, will all vary. Internally the form of the core and the color and texture of the fruit are characteristics unique to each apple.

The Importance of Chromosome Numbers

The way an apple tree produces fruit varies. A few rare individuals can fertilize themselves. These are called self-pollinating or self-fruiting trees. Some are partially self-fruiting and so will set (i.e., produce) some fruit by themselves, but they will also benefit greatly from another tree's pollen. Most apple blossoms require the pollen of another tree to set fruit. Orchards should contain two or more cultivars that flower at the same time to ensure proper pollination.

Apple trees can also have different sets of chromosomes, the structures that contain the genetic codes that will determine an apple's characteristics. Those that have two sets of chromosomes are called diploids. Another smaller group called polyploids has three or more sets and requires the pollen of two or more apple trees to become fertilized.

Polyploids usually have fewer fertile flowers, but this is not a great disadvantage. Most apple trees produce far more flowers than they can possibly use. Most polyploids also have little viable pollen, which means such apple trees cannot be relied on as pollinators for other apples. Among polyploid apples the most common group is known as triploids. These trees carry one extra set of chromosomes. Triploids require two diploid pollen parents (two different cultivars) for pollination. The first diploid pollen donor to arrive at the ovary will fertilize the extra set of chromosomes, but not the diploid set. If a second pollen donor arrives it will pollinate the diploid set, and pollination will be complete. In orchards containing triploids, such as Rhode Island Greening, Golden Delicious or Bramley, there must be at least two other diploid pollinators that flower around the same time to be sure of good pollination and fruit.

These differences are important to know when you begin growing apples, as you will need to have compatible cultivars that will pollinate each other. Luckily, there are lists you can obtain that will let you know if your apple tree is a diploid or a triploid. You will also need trees that flower at the same time, or at least have overlapping flowering. If they do not flower together there will be no fertilization. In general, early-season apples blossom before later-season apples.

While it may sound complicated, the reality is that if you have an orchard with several different cultivars you will probably have no problems getting fruit. If your orchard is isolated, proper pairing of cultivars is a bit more important. In most situations, however, there are likely other apple trees in your vicinity whose pollen will arrive on the legs of pollinators such as bees, which often fly long distances when gathering pollen.

Frost and Flowering

Frost is the nemesis of all apple growers. Late frosts in the spring can be devastating if the trees are in bloom. Once temperatures reach 32°F (0°C) the cells in the delicate parts of the flower rupture and wither, making pollination impossible. Siting trees so that they are in the most advantageous position during frosts can mean a difference of as little as half a degree, which can also mean the difference between freezing and not. (See Chapter 3 for more about siting your trees.)

Cold air drains from higher areas to lower areas on still spring nights and will pool in the lowest areas. The most effective way to stop flowers from freezing is to plant your trees on higher ground or on the slopes where air is passing through the orchard.

Commercial growing areas are nearly always located beside large bodies of water, such as large lakes, rivers or oceans. Some are on hillsides where air drainage is good. Though not everyone is so lucky, you can site your trees to take advantage of your terrain.

The flower's promise of fruit can be ended ▸ when temperatures dip below freezing.

Another possibility is planting trees on the north side of a hill, where the cooler temperatures delay flowering. Such sites are not as amenable to long-season cultivars, but they may delay flowering long enough to miss the last spring frost.

I have found that pruning later in spring, but before flowering, can delay flowering by a few days.

The Pollinators

Without insect pollinators, the bounty of fruit trees would become a paucity. It is the movement of pollinating insects, particularly bees, that transfers pollen from flower to flower. When apple trees are in flower, bees will be moving from one tree to another, gathering pollen, which is placed into sacs on their legs. The pollen ends up all over the bee, and as the bee lands to gather pollen from a new flower many of the pollen grains from the previous flowers will fall off. Some will land on the fertile pistils, the organs of the flower that trap the pollen with a sticky material. Once on the pistil, the pollen will germinate and grow with incredible rapidity down the style to the ovary, where fertilization takes place.

Conditions that are unfavorable to bees during flowering are unfavorable to a good fruit set. Rain, cold and strong winds during flowering can reduce the crop dramatically.

Recently, problems have developed that have affected the survival of honeybee colonies. Varroa mites, viruses and other poorly understood factors have reduced or decimated many colonies. While wild bees can take up much of

Not all polllinators are bees. Here a hoverflly unwittingly acts as a pollinator.

Tubes with an inside diameter of 0.4 in. (1 cm) are used here to attract mason bees. The tubes should be sealed at the far end and removable for cleaning each year.

the slack, large orchards are often unsatisfactory areas for wild bee colonies, so if honeybee colonies cannot be brought into such orchards, it will be potentially devastating to the industry.

Few small orchards or home growers will consider bringing in bee colonies. That means you have to rely on wild bees and other pollinators. To encourage these, you should try to create conditions that will attract and maintain them. Diversity of plant material is a key factor. Many pollinators harvest pollen and nectar from a wide range of plants. The more different plants you can have growing near your trees the better the chances of having pollinators. If you are growing apples in a suburban or urban setting, this can be a challenge as mowed lawns and concrete are deserts for pollinators. In such settings you can encourage pollinators by planting high-nectar flowers such as butterfly weed, sedums, lilac, bee balm and weigela. Though your neighbors may frown upon it, leaving the grasses and wildflowers to grow under your small orchard can help immensely. Simply mow them at the end of the fall. You can encourage such pollinators as the mason bee and the leaf cutter bee by buying or building specialized "homes" for them.

3 Planning Out an Orchard

The architecture of an orchard gives us row on row of trees, planted equidistant to give them equal access to the sun. In bloom, they are enchanting places to walk, surrounded by the sweetened airs of Earth as a silent rain of white and pinkish petals falls to the ground. During harvest they are bursting with juicy orbs just begging to be plucked.

Whether you are planting a few trees or a few thousand there are certain guidelines that should be considered when establishing an orchard, which include site, drainage and spacing.

◂ Walking through an orchard in bloom is one of the greatest joys of growing apples.

Site

The grower who attempts to cultivate fruit in northern areas must come to terms with certain limits if their venture is to prove successful. This includes not being able to grow the fruits you might desire; but bear in mind that growers living in warmer climes are limited as well and may crave the fruits you grow with ease — gardeners living among mangoes might crave raspberries. It seems in our nature to desire what we are told we cannot have. There is

The apple's hardiness makes it a great northern fruit. Some apples cling to their stems well into the winter.

always merit and reward in testing the limits of plants, but understanding the realistic boundaries will keep disappointments to a minimum. That being said, northern gardens have many choices, and more than ever before.

Winter survival is not just a tale of temperatures. The flow of weather, when low temperatures occur, how quickly temperatures fall, the velocity of the wind and other factors play an important role in deciding whether all, part or none of the buds on a plant will leaf out in spring. On a given site some of these factors can be influenced by our work, but others, most essentially the actual temperature, will be determined by where you are on Earth's surface. As you proceed northward or skyward, you usually experience colder minimum temperatures.

Every plant has a limit beyond which its cells die. For cells plump with water the freezing point is the critical limit. When water freezes, it crystallizes and expands, piercing and rupturing the cell walls. Plants with an ability to withstand colder temperatures respond to the decreasing daylight hours and declining temperatures by lowering the water content in their cells, redistributing the water to spaces between the cells or into the root system. Very hardy plants can remove virtually all water within their cells, leaving only a thin film around each cell's vital parts or organelles that can remain flexible (an oddity of water), even at extremely low temperatures.

The minimum temperature any plant can endure can be affected, often dramatically, by the timing of the low temperatures. Plants develop deeper dormancy as winter approaches. If the progression toward that state is gradual, the plant will successfully enter into a deep dormancy. If low temperatures are experienced early in the fall, when the plant has not yet reached deep dormancy, cold that would not harm more dormant plants can injure or kill cells.

Midwinter survival can be influenced by prolonged thaws that draw plants out of dormancy and are followed by a quick descent to cold temperatures again. When a plant is drawn out of dormancy it will not return to as deep a state of dormancy. Unfortunately, even the best of sites cannot counter such random events, so an ideal choice of cultivars will be your best defense.

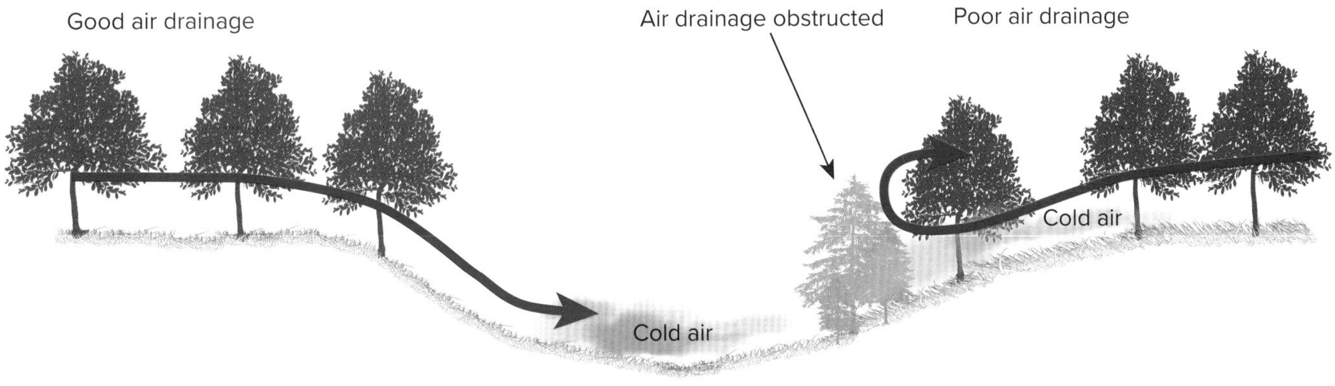

The site you choose for your apple orchard will be critical to its future success or failure, and there are several aspects you must consider. First, air drainage is important for preventing damage from late-spring frosts. If temperatures dip below the freezing point during flowering, there will be little or no crop. Though this may not be a problem for the home grower, it is a financial disaster for the commercial grower.

Orchards placed where cold air settles on still nights will experience more frosts. Early-fall frosts can damage tissues that have not had a chance to harden for winter. Late-spring frosts can damage emerging leaves and the flowers so vital to fruit production. It is no surprise that commercial orchards are usually located near large bodies of water that moderate night temperatures and/or on hillsides, where air is moving during nights when the difference of a few feet in elevation can mean the difference between freezing or not.

Many large orchards install immense fans that mix cold air at the ground level with the warmer air above the orchard, lessening the chance of frost damage to the flowers.

If possible, your trees should be sited at the top of a hill or on a side hill where heavy cold air falls away toward lower areas. If trees or other obstructions at the bottom of the orchard trap cold air, it is best to clear these to allow the cold air to escape to even lower areas. As touched on in the previous chapter, some growers plant their orchards on the north face of a slope. This aspect is cooler, and can delay flowering by several days, which may prevent injury to blossoms from a late frost.

Sites facing the southwest can experience damage aptly named "southwest injury." At sunset on sunny days, most often in late winter when the sun is stronger, the tree's dark bark is heated until it thaws, and water is absorbed back into the cells. When the sun sets the temperature drops abruptly and the cells freeze, causing the bark to pop open. The injury shows up as a vertical split precisely aligned to the spot where the sun has set. If this is a problem, the solution can be as simple as painting the trunks white to

Well-drained, rich soil is absolutely critical for good growth no matter the density of the orchard.

reflect the sun or shading the trunk with a board or plants. Luckily, apple trees are less subject to this injury than many other trees.

Cold dry winds can harm plants by desiccating their stems. This can often be ameliorated by planting or using existing windbreaks.

Drainage

Soil texture is among the most critical criteria for an orchard. Apple trees prefer well-drained open soils. Although apple rootstocks differ in how they tolerate poor drainage, any successful orchard will have either good natural drainage or will have some form of drainage created. This might involve installing a perforated drainpipe below the roots to direct groundwater toward a lower area or creating built-up areas called berms on which the trees are planted that alternate with lower swales to carry away excess water. The goal is to make sure water is never allowed to saturate the root zone.

Sandy, silty and gravelly loam soils are ideal for apples, though if the soils are so open as to be dry, some form of irrigation will have to be installed. A well-drained clay loam is acceptable, though subsurface drainage in such soils is a wise investment to protect against the possibility of prolonged wet conditions, particularly in winter. Saturated soils in winter are an invitation to disaster.

I go into further detail about the importance of soil in the next chapter, as it is a subject that deserves to be explored in greater depth.

Spacing

The spacing of trees depends on the rootstock, the cultivar and the richness of the soil and availability of water. All these factors should be considered before deciding on your spacing. Consulting with local growers and apple specialists is an excellent way to avoid future frustration.

Rootstocks vary considerably in the amount of size control they impart to the cultivar. The natural vigor of the apple will also influence the ultimate size of the tree. A rich soil will grow a larger tree, and the capacity of a soil to provide water throughout the growing season will also help influence size.

Lastly you should consider how you will manage the orchard. The size of equipment used to tend and harvest your orchard needs to be taken into account.

A one- or two-year-old tree is small, and it might be tempting to plant the trees close together, but remember that small trees grow into larger trees. Seedling rootstocks can form very large trees. Choose your spacing with the future size of the trees in mind. Pruning can help maintain size, but the initial spacing is the most important factor for access to sun, air movement and your ability to move through the orchard.

Many nurseries and resources have charts that can help you determine how many trees you will need to fill a given area for the spacing you choose.

High-Density Orchards

The move toward high-density orchards using trees grafted on dwarfing rootstocks has intensified over the past several decades. Virtually all new orchards are planted in rows with the trees averaging 2 to 3 ft. (0.6 to 1 m) apart, occasionally closer. Initial capital costs for the support system and trees are much higher than for a lower-density orchard. The grower must provide a strong and long-lasting system of posts and wires to support the trees for their entire lives. If all goes according to plan, the orchard will start producing revenue earlier and the production will be higher, especially for high-grade fruit.

Large-scale plantings use a vertical pruning machine to maintain the narrow hedges. The apples are picked using narrow tractors with specialized equipment and often with pickers at two different heights standing on platforms attached to the machine. No doubt there will soon be robots that will pick the fruit without the labor of humans.

The math has been favorable for high-density growing techniques. Quality is easier to maintain, as the branches are easily accessible from the ground or picking machines. It is also safer and more comfortable for growers and pickers, as they are not required to climb tall trees.

However, problems can develop in high-density orchards. If the pruning equipment is not kept sharp and clean, infections to the trees can develop and be spread by the machine. Although the sickle bar–type pruners are efficient at keeping the width of the hedge consistent, detail pruning must still be practiced to

keep the tree open to sunlight and to prune lateral branches and spurs for optimal production and size. (See Chapter 5 for details on pruning techniques used in high-density orchards.)

Relying on one type of rootstock can be disastrous if a condition or disease that affects the roots strikes the orchard, though this is true in any density of planting.

Today the bulk of our apples come from high-density orchards. It is another move toward the industrialization of the agricultural sector, though the fiscal argument for such orchards is strong. If well grown, these orchards produce more quality fruit per acre and, over time, will yield a higher profit margin for the grower. However, for many growers the lower-density orchard will remain the preferred system. It allows the tree to grow into its natural form and creates a space many find personally satisfying. It also frees the grower from the maintenance of posts and wires, as most trees in lower-density orchards are grown on self-supporting rootstocks.

◂ High-density plantings have completely changed the look of commercial orchards.

4 Soil

The Unseen Ecosystem

We owe our lives to those who toil in dark and unsung places, roaming amid the scraps of rotting leaves and grains of sand, eating remnants left by those that lived before. For it is death and decay that fosters growth and, through that growth, the fruits we so desire. All too often we ignore decay, focusing only on the prettier and tastier aspects of gardening and farming. However, a clear understanding of how things rot will lead to both a deeper insight into how to enhance growth and hopefully a better appreciation for the importance of soil.

Soil is, in large measure, the remains of mountains ground and dispersed by gravity, water, ice, rain, wind and volcanic fury. These bits of rock provide but a matrix for a web of life we are only beginning to comprehend and one deserving of awe for both its complexity and centrality to our existence. While the particular mix of rock particles can have an important impact on the fertility of a soil, it is the life among those particles, both animal and vegetable, that have the greatest influence on the ability of a soil to provide the nutrients necessary to grow healthy plants.

◂ A soil rich in organics is one of the keys to a productive orchard. Compost is a great soil amendment, whether added to the planting hole or spread as the orchard matures.

Animals that live in the soil, such as eathworms, aid in the cycling of nutrients and help improve the soil's drainage, structure and general productivity.

Those who live in northern latitudes, or at high altitudes, inhabit lands that were most likely scraped by glacial ice that removed any soil accumulated in warmer eras. The history of most northern soils begins only 10,000 years ago, when the glaciers retreated toward the poles and the first lichens, mosses and algae began repopulating a landscape nearly devoid of organic matter. When the slow process of organic accumulation began, plants such as birch, whose seedlings are capable of surviving in rock scree, created new forests along with other pioneer species, and an annual rain of leaves began the slow (at least from our perspective) process of soil development.

As more plant and fungus species returned to the land, animal life returned as well. These animals ate seeds, leaves, stems, roots and other animals. A barely noticeable accumulation of bird droppings, millipede poop and such aided the decomposition of leaves and stems by providing bacteria and other soil life with higher levels of soluble elements, such as nitrogen, phosphorus and potassium. A balance was created between the amount of organic material waiting to be digested and the number of life-forms available to do the job, whose numbers were in turn determined by the availability of those crucial elements. What tips this balance toward the creation of soil is the energy of the sun, captured by plant cells and converted into another generation of leaves, stems and roots.

Proteins in dead plants are broken down when the plants are digested by animals. The unused nutrients are excreted, resulting in the deposition of soluble nitrogen. When animals die, more soluble nitrogen is added. Although plants use many elements and compounds for growth, nitrogen plays a vital role because plants cannot directly absorb nitrogen from the air, where it is abundant. They can only absorb it through their roots when it is present in the soil water (or, in a few cases, when living in cooperation with nitrogen-fixing bacteria).

When nitrogen stimulates plant and animal life, more growth and higher levels of reproduction begin to speed the process of organic decomposition. The process of soil building is, for our purposes, very slow. After 10,000 years of growth and decay, most upland sites might have a layer of soil 1 to 3 ft. (30 to 100 cm) thick. This is a far cry from the voluminous depths of great soils in river valleys, deltas and plains, where the deposition of soil-building materials is more rapid, and perhaps where the process may not have been interrupted by glacial scraping.

The reason for the slow accumulation is due to the nature of organics. Though the leaves, stems and roots of plants seem to have an appreciable mass, a giant tree becomes a few pinches of dust when fully broken down. Even pinches, however, accumulate, and eventually soils develop deposits of organics that break down over millennia rather than years. These stable carbon compounds, called humus, in combination with organics that are in the process of decomposition, give the upper layer of soil, the topsoil, its dark color and spongy texture.

The opening of soils and the subsequent erosion of these delicate surface layers of topsoil are tragic because this rich productive layer is essentially irreplaceable. A soil that took 10,000 years to create can be washed away in a single downpour. Most of the best soils on Earth have already been lost to wind and rain because of the removal of the plants that mechanically hold such soils together with their roots and the systems of fungi associated with those roots. Stripping vegetation from the earth leaves the light organic matter vulnerable to the forces of erosion. The degradation of soils around the world represents a loss of potential that is incalculable.

During the past century, scientists were able to reduce the nutritional needs of plants to a series of nutrients, such as those mentioned above. While this elucidated the elements necessary for growth, the importance of soil biology, structure and rejuvenation were often disregarded. Rather than rely on older methods of soil maintenance, soils were, and still are, regarded as merely the substrate on which the farmer or gardener can incorporate the necessary nutrients, usually delivered as soluble salts. The results, often called the Green Revolution, created incredible short-term increases in yields. Over time, however, problems began to surface. Without crop rotation and cover crops, open fields became subject to erosion, often on a massive scale. The tilth or texture of soils breaks down as the organic matter is decomposed by the incorporation of nitrogen from fertilizers and the increased supply of oxygen created by cultivation. What organic matter existed in the soil is biologically burned up in the slow "fire" of rot, ending up in the atmosphere as carbon dioxide.

Organic matter, once replaced by manures, cover crops and fallowing, became depleted as farms relied more on soluble fertilizers. Increasing scales of production necessitated larger fields. This led to the use of larger equipment, which was needed to accomplish the farming chores on so much land. The hedgerows that grew between small fields were cut to accommodate the machinery. Wind became a larger factor in soil erosion and where there was elevation, water erosion. Wind and flood events can remove large amounts of soil, and the lightest portions of the soil go first. The organic portion of the soil is the most valuable. A single rainstorm falling on a freshly harrowed field atop a steep hillside could easily remove most of a soil's potential for growth. Unprotected soil is a most fragile thing.

What was poorly understood in the heady days of deconstructive agriculture was how soil naturally provides nutrients to plants. Continuous cultivation and the annual application of soluble salts, often highly acidic in nature, destroy much of the microbial and fungal growth that associates with plant roots and is essential to nutrient absorption.

It was generally assumed that plants were simply delvers, intruders if you will, into the mysterious world of the soil where they absorbed the water and nutrients they needed. As we find out more about the relationship between plants and soil, we are discovering the explanation is far more complex. Over the ages, plants have developed unique and symbiotic (mutually beneficial) relationships with many organisms. One of the most interesting of these is the association between plant roots and several species of fungi that are grouped together as mycorrhizae.

A plant root has a certain amount of surface area from which its specialized outer cells can absorb nutrients and water. If the soil dries, the plant has only so much soil it can tap into for water. However, mycorrhizal fungi in the soil connect with the plant roots and, with the aid of sugars supplied by the plant, spread root-like structures called mycelia to delve deep and wide, allowing the plant to draw water and nutrients from an astonishing amount of surface area. In fact, plants grow specialized cavities in their roots to accept the mycelia of these fungi. The plant will send nearly one-quarter of its food resources to the roots to provide sugars to the mycorrhizae in return for the water and nutrients. It is a good arrangement for both plant and fungi. One of the principal reasons transplanted trees recover slowly is the lack of a mycorrhizal web to provide water and nutrients. There are now commercial preparations that can be added at planting that will more quickly restore these mycorrhizal systems.

The cultivation of soil wreaks havoc on these

Beneath the soil's surface is a complex world inhabited by bacteria, insects, animals and fungal networks that deliver nutrients to the trees.

relationships. Massive webs of fungal growth are torn apart and exposed to air. Continued cultivation can result in the substantial reduction, or even extinction, of these vital soil fungi. Once gone, plants can never reach the potential possible in a soil where these relationships are present. The crux of the dilemma in present-day farming and gardening is how to have the benefits of cultivation (aeration, weed suppression and decomposition) without the detriments of soil disruption and degradation.

At one time many apple growers believed that continual cultivation was beneficial to the trees. If your soil was deep and the land flat you got away with it for a while, but soon the tilth of the ground was destroyed. It was discovered that apples lend themselves to be grown with permanent ground covers, such as various grasses, clovers and other plants, which are usually cut at least once a year. Such a system means that the soil systems are minimally disturbed, and this is a benefit to the grower if managed correctly. Most high-intensity growing systems use herbicides to kill the competing plants under the tree but leave a permanent cover between the rows to prevent erosion and make it easier for vehicles to access. The organic grower can accomplish the same with mulches in the row and/or permanent cover crops between the rows.

Acidic | Neutral | Alkaline

0 1 2 3 4 5 6 7 8 9 10 11 12 13 14

Soil Acidity: The pH Factor

The pH scale is a way to measure the acidity level of soils. On this scale 0 represents the highest level of acidity, and 14 the highest level of alkalinity. Pure water has a pH of 7 and is considered neutral. Hydrochloric acid has a pH of 0, and lye or drain cleaner has a pH of 14. For each decrease of one point on the scale, the acidity of a soil increases 10 times; therefore, the difference between a pH of 7 and a pH of 4 (10 x 10 x 10) represents a thousandfold increase in acidity.

Measuring the pH level of soil is vital in any sector of agriculture, and no less so when growing apples. The soil's pH has tremendous implications for a tree's ability to absorb nutrients. In acidic soils many nutrients, such as calcium, phosphorus and magnesium, are chemically bound to other elements, making them insoluble and unavailable to the tree. Likewise, very alkaline soils bind elements.

In northern areas soils tend to be acidic. If a soil's pH is below 6.3 an apple tree will not be able to access the levels of nutrients necessary for optimal growth and fruit quality. Such soils should be brought toward neutral.

The most common and inexpensive neutralizing agent for acidic soil is agricultural limestone, also called agricultural lime. This product is limestone rock that has been heated so that the chemical water contained within expands and breaks apart its internal structure, making it easier to pulverize the rock to a fine powder. Powdering the rock increases the surface area of a given volume exponentially and greatly accelerates the weathering of the calcium-rich rock. As the calcium dissolves in the soil's water, it bonds with free hydrogen ions, which are abundant in acidic soils, and the soil becomes less acidic.

Limestone usually contains varying amounts of magnesium as well. Those with high levels are called dolomitic limestones. If your soil has low levels of magnesium you should seek out dolomitic limestone.

There are other acid-neutralizing agents available, such as phosphate rock and wood ash. Phosphate rock is ground calcium phosphate. This product has only one-half the neutralizing capabilities of lime but does contain high levels of phosphorus. (Phosphorus is generally present in most soils, but the pH must be adjusted to make it available to the tree.) Perhaps the most pressing issue with phosphate rock is the rapidly diminishing supplies of this product. There are only a few known large deposits in the world, and these are quickly being depleted.

Wood ash is commonly available in rural areas where wood is used for heating. It is a more powerful acid-neutralizing agent than lime and should be used cautiously. It does have significant levels of phosphorus, potassium, calcium, magnesium, zinc, boron and copper, and it can be useful if moderate amounts are incorporated around your trees.

The amount of neutralizing agents needed to

affect the proper change in pH level can be complicated by such factors as the amount of aluminum and iron in the soil. It is highly advisable to make use of provincial or state agricultural agencies that perform soil tests. By sending in samples of your orchard's soil, you will be able to get the exact quantity of neutralizing agent necessary to create the ideal pH level.

Neutralizing the soil can take some time. It may take several years to reach the pH you desire. Eventually the pH level will plateau. After several years, because rainwater is acidic, the calcium and other soluble minerals will gradually dissolve, and their neutralizing effect will decrease. As a general rule, for areas with naturally acidic soil a light application every six years or so is recommended to keep the pH level where you want it.

A few northern areas have alkaline soils. Soils above 7.3, especially those with carbonates, are difficult to neutralize and are not good for growing apples. If the soil pH is below 7.3 you may be able to lower it by adding elemental sulfur. The soil bacteria convert the sulfur into sulfuric acid, which lowers the pH. Again, it is advisable to have your soil tested to find out the proper amounts of sulfur needed to lower the pH to the desired level.

The importance of neutral soil pH in an orchard cannot be overstated. Though an apple tree can survive in soils with lower pH levels, you will never achieve optimal growth and production until the soil is neutralized. It is vital to the success of your endeavors. Adjusting the pH of the soil allows the trees to access the soil's nutrients. Every grower of apples needs to be aware of this.

Providing Macronutrients

Besides the pH of your soil, the levels of nutrients found in your soil will also determine the health and vigor of your trees. Below are the important nutrients that are needed to create great apples and apple trees.

Nitrogen

Nitrogen has been called the most important element for growth. In truth, there is no "most important" element. Every element necessary for growth must be available to grow a healthy tree. Nitrogen is critical for shoot and leaf growth, but it is often in short supply and thus is given priority in most fertilization programs. Although 78 percent of the air around us is nitrogen, plants cannot access it directly. With few exceptions, they rely on nitrogen compounds created by the decomposition of organic matter and the waste products of soil creatures.

The so-called chemical fertilizers, such as ammonium nitrate, have nitrogen compounds that, when dissolved in water, can provide an immediate source of soluble nitrogen, but these materials have several drawbacks. These compounds are highly acidic and in excess can destroy much of the soil life that helps provide nitrogen naturally. In addition, their production requires vast amounts of energy, negating much of the caloric energy gained by the food they help produce.

Nitrogen can be provided naturally by adding compost as well as using animal by-products such as manures, blood meal, fish meal, feather meal and bone meal. Some

Leguminous crops, such as clover, alfalfa and bird's foot trefoil, capture nitrogen from the air. The nitrogen eventually becomes available to the apple's roots.

high-protein vegetable products are also good sources of nitrogen. These include soy meal, cottonseed meal and alfalfa meal. Some organic standards allow the use of moderate amounts of nitrate of soda, a naturally occurring material mined in the dry deserts of Chile and Peru.

Fresh manures are not advised for orchards as they can contain bacteria, such as *E. coli*, that may find their way onto the surfaces of the apples. This is especially pertinent where apples that drop to the ground are used for non-pasteurized cider. Fresh manures, however, are invaluable as sources of nitrogen when used to create compost.

One of the most efficient methods of introducing nitrogen into the orchard is to seed crops such as alfalfa, clover and bird's foot trefoil. Such plants are called legumes, and these form associations with specialized bacteria on their roots. The bacteria capture atmospheric nitrogen and transfer it to the plant in exchange for sugars. A vigorous stand of legumes disked into the soil in spring can provide 11 to 22 pounds (5 to 10 kg) of nitrogen per acre, of which half will become available during decomposition. This will provide sufficient nitrogen for most orchards. A permanent stand of legumes such as clover will provide a slow release of nitrogen as the plants die as a result of natural competition.

Material	Percentage of Nitrogen Found in Material
Fresh dairy manure	0.5%
Fresh chicken manure	1.5%
Poultry manure compost	4%
Finished compost	1.2%
Legume hay	2.5%
Grass clippings	1.2%
Alfalfa meal	2.7%
Soybean meal	6%
Blood meal	12%
Cottonseed meal	6%
Crab meal	5%
Fish meal	9%
Feather meal	10%
Nitrate of soda	16%

Source: Vern Grubinger, "Sources of Nitrogen for Organic Farms," University of Vermont

Organic sources (with the exception of nitrate of soda) do not provide nitrogen directly, as soluble fertilizers do, but rather provide a source of food for bacteria and other soil organisms. When these are introduced, populations of soil organisms explode. Newly incorporated nitrogen-rich materials can cause populations of soil bacteria to quadruple every hour. Once ingested, they pass through the bodies of these creatures and their fecal matter becomes a soluble source of nitrogen and other elements plants can absorb. This cycle tends to keep the nitrogen in the upper reaches of the soil, making it available for several years.

Phosphorous

Phosphorous is an essential nutrient that is important for the formation of seed and critical to root development. When the pH is adjusted to optimal levels, most soils have adequate amounts of phosphorus. When phosphorus levels are inadequate many growers rely on various soluble forms of phosphorus, such as superphosphate. Organic growers can use sources such as phosphate rock or bone meal.

Phosphorus is most crucial during a tree's early life, when root development is rapid. Incorporating a few handfuls of bone meal when planting will often be sufficient for a tree's needs.

Potassium

Potassium helps produce quality fruit. Low levels may result in small fruit size, low sugar content and poor storability. In soils deficient in potassium, the tree may show symptoms such as small bluish leaves with dried yellow edges. Most soils will have adequate levels of potassium if the soil's pH is between 6 and 7.

The most common commercial source of potassium is muriate of potash. This is a salt form of potassium that can harm soil life, particularly if used in high concentrations. Organic growers generally rely on Sul-Po-Mag or sulfate of potash-magnesia (22 percent potash), as it is kinder to soil organisms. As the name implies, it also contains high levels of sulfur and magnesium. If available, wood ash is an excellent source of potassium (6 percent), as well as calcium and a host of micronutrients.

Calcium

Calcium is a primary constituent of cell walls. It moves more slowly through a tree's vascular system than most elements and a deficiency can cause several disorders in apples. The most common of these is bitter pit. The key symptom of bitter pit is a shallow pit on the apple's surface resembling a bruise that looks as if it was caused by a pointed object. The inside of the pit will be gray or brown, and underneath the pit the flesh will be spongy, dry and brown. Bitter pit may not show on the fruit until after harvest.

Watercore is a similar disorder that is affected by calcium levels in the fruit. Both this and bitter pit are discussed further in the physiological disorders section (see pages 92–94).

Although other factors may be involved in creating these disorders, the most important part of any long-term solution is to ensure your pH level is near neutral. Such soils usually contain good levels of calcium. When the soil's pH level is between 6.5 and 7, there is generally enough calcium available.

Magnesium

Magnesium is vital in the formation of chlorophyll, which makes plants appear green and through which photosynthesis takes place. The ability of a soil to provide adequate magnesium for an apple tree depends on the soil's pH. In acidic soils, the element is most often bound up in molecules that the plant cannot uptake.

Bitter pit is due to a lack of calcium during the fruit's development and shows up as the apple matures.

In neutral soils there are more free ions in the soil's water that the plant absorbs. A lack of magnesium will show up as yellowing in the leaves, particularly older leaves and/or those exposed to more light. Soils deficient in magnesium can benefit from a high-magnesium (dolomitic) limestone, though this is not recommended for general use, as too much magnesium can influence how much calcium the tree absorbs.

Sulfur

Sulfur has recently been included as one of the macronutrients necessary for good plant growth. It is important in the production of certain amino acids for protein formation and essential for creating enzymes that are used to form chlorophyll. Most sulfur is found in a soil's organic matter, which must be broken down so that the absorbable form of sulfur becomes available. Once inside the plant it is not mobile, so if there is a lack of sulfur it will become noticeable in younger leaves that are small and that show yellowing between the veins.

Providing Micronutrients

Micronutrients are equally important to plant growth but are usually found in quantities sufficient for a tree's needs. These include boron, chlorine, manganese, iron, nickel, copper, zinc and molybdenum. However, there are soils that are deficient in these nutrients. A good example is boron. Although needed in exceedingly small quantities, without any boron an apple, or most any other plant, will not grow well. Before planting an orchard, it is wise to have your soil tested for every important element. A good quantity of organic matter in the soil will generally provide the micronutrients needed for optimal growth — a great case for using compost.

5 Planting and Maintenance

Planting

A tree can arrive at your yard or orchard as either a bare-root plant or a potted plant. This section advises on the best planting methods for both.

Bare-Root Trees

Because the roots of a bare-root plant have no covering, it is essential the roots are kept moist, preferably with a moisture-retaining material such as fine-wood fiber or damp shredded paper. The roots should be kept with this material in an airproof bag and not removed until the tree is being planted. Never leave the roots exposed to the sun and wind, even for a few minutes, especially when it is particularly warm and/or windy.

Prepare the hole to a depth equal to the distance between the lowest root and the crown, the area where the root system meets the trunk. If the crown is buried, there is a greater possibility for damage due to crown rot, a fungal disease caused by *Phytophthora cactorum*. This is most critical when planting in soils that are slow to drain. The graft union should be 4 to 6 in. (10 to 15 cm) above the crown.

◂ Once planted, an orchard requires consistent and thoughtful maintenance to produce quality fruit.

Bare-root trees should have their roots spread evenly in the planting hole.

Make the hole twice as wide as the root system. Remember width is more important than depth, as feeder roots are more apt to roam near the surface, where food and oxygen are abundant.

If your soil is clay it is best to use the soil from the hole amended with up to 20 percent organic matter, such as compost, and organic fertilizers, such as fish, blood or alfalfa meal. If the soil pH is low, add lime into the soil and spread it in a wide area around the tree. Sprinkling a commercial mycorrhizal fungi preparation on the roots is highly recommended as well (see pages 42–43 on the importance of mycorrhizae).

Once you have established the crown level (a shovel handle across the hole works well), work the soil around the roots, being careful not to leave air pockets that may injure the delicate outer tissues of the roots, where water and nutrient absorption occurs. Once the hole is filled in, tamp the soil lightly with your feet.

It is helpful to leave the surface slightly dished or to create a temporary ring of higher soil so that water is directed down toward the roots only. Water the tree deeply twice to settle the soil around the roots and to make sure the roots are surrounded by moist soil. For the first few months, be sure to maintain a schedule of watering. A good deep watering once a week should suffice in most instances. In poor-draining soils adjust to fewer waterings so that the soil remains moist but never waterlogged.

When planting numerous trees an auger can reduce labor immeasurably; however, be aware that in soils with high clay content the auger can cause the particles at the edge of the hole to smear together, creating a glaze that resists water movement. If this happens, be sure to score the edge of the hole with a shovel to improve drainage through this denser layer.

Depending on the size of the tree it is wise to place one or two stakes to stabilize the tree for the first year. Use flexible ties such as cloth strips so that the tree is kept vertical but can still flex a bit. Wind stimulates the production of stress wood in the trunk and makes it a stronger tree. Remove the stakes after one year. Plantings of apples on dwarfing rootstocks are generally tied to wires strung between posts and so that they may not need individual stakes, though some growers still stake the trunk as well as tie branches to the wires.

Crown Gall Prevention

Crown gall is a bacterial disease caused by the soil-borne *Agrobacterium tumefaciens*. The bacteria cause the tree to form large, irregular galls, typically on the roots or close to the soil line. The bacterium can enter the tree by mechanical injury, insects or planting damage. Crown gall can reduce the future vigor of the tree and at its worst may kill a young tree by girdling the stem.

Another bacterium called *Agrobacterium radiobacter* can mitigate the risk of a crown gall infection on the roots. Right before planting, dip the roots of the tree in a solution containing *Agrobacterium radiobacter*. Enough bacteria will adhere to the surfaces of the roots to protect them against the crown gall bacteria in the soil. (Read more about this on page 128.)

Be sure to tease out any circling roots before placing a potted tree in the planting hole.

Potted Trees

Planting a potted tree differs little from planting a bare-root tree except that the roots are contained within the limitations of the pot. When you remove the tree from its container you will often find roots coiling around the outside of the root ball. If possible, tease these out so that they can be spread outward in the planting hole. If left coiled, such roots can eventually constrict the expanding trunk and, in some instances, can even shorten the life of the tree. In severe cases where the roots are so tight they cannot be pulled out, it is a good idea to take a knife or pruners and make several vertical cuts into the root ball about 1 in. (2.5 cm) deep. This will force the creation of new roots that will delve into the surrounding soil.

Once you have positioned the tree and spread the roots as best you can, work the soil around the roots and root ball. After the hole is filled, tamp lightly with your feet, and then water and stake as described above.

Replant Problems

When older apple trees are removed and new ones planted, the new trees can become stunted or even die. A great deal of research has been done on this problem. Replant diseases usually involve soil-borne fungal pathogens such as *Pythium*, *Rhizoctonia* and *Cylindrocarpon*, but there is also a culprit called the root lesion nematode that plays a part. These microscopic worm-like creatures live in soil and within root tissues. Their feeding will often introduce the fungal pathogens into the root system and can cause distress or even the death of the tree.

A larger tree can generally handle a high population of root-feeding nematodes. However, when the old tree is removed this population is left without food. When a small new tree is planted soon after removing the old tree, this hungry population will find and feed on the new roots in numbers that may seriously harm the tree. Because the young tree has not developed its defenses, including relationships with various soil fungi, it cannot repair itself quickly enough. It loses the ability to absorb nutrients and water, and suffers loss of vigor.

The best defense is to create conditions that prevent the root-feeding nematode population from proliferating. When replanting, it is advisable to delay planting a new tree for at least one year or to plant between the original planting sites where nematode populations will be smaller. Planting in ground where apple trees have not been grown before is ideal.

The soil pH should be maintained at 6.5 to 7. The soil should also be adequately fertile. Phosphorus deficiency, in particular, can be a problem. Applying composts and soil amendments can create an active soil with a diverse population of soil organisms, and diverse soils help prevent overpopulations of any individual species.

Weed Control

When planting new trees, it is advisable to keep grass and weeds away for the first year or two. Competition for nutrients and water can have a marked effect on the growth of small trees. Commercial growers often resort to herbicides as a cost-effective way to manage weeds. Organic growers should consider mulching. Mulches might include rotted bark or sawdust, straw, grain hulls or even several layers of thin flat stones. Spreading newspaper or cardboard beneath the mulches is an excellent way to ensure perennial roots do not re-establish. Remember that high-carbon mulches, such as fresh sawdust or bark, will reduce nitrogen levels when bacteria begin the breakdown process, so it is wise to incorporate a nitrogen-rich amendment to the soil before laying down such mulches.

Mulching can provide future organic material while keeping weeds in check. Using cover crops between rows helps promote good soil health and prevents erosion on slopes.

Some growers prefer to maintain a weed-free area around their trees as they age, but most growers use grass as a permanent soil cover in their orchards. Grass ensures there will be no erosion on slopes and provides good winter protection for the tree's roots because the layer of grass and grass roots acts as insulation. Upkeep of grass is time and cost effective, as one or two cuttings a year is usually sufficient for maintenance.

In larger orchards swing arm mowers can be a great time-saver. They can reach underneath a tree's canopy and cut right up to the tree. They also have a retractable arm that automatically senses the trunk and pulls back to prevent damage to the trunk. In smaller orchards handheld mowing devices can be used, but whipper snippers and mowers should *never* be allowed to strike the trunks. If the cambium layer beneath the bark is cut, there can be irreparable damage to the tree. This is particularly pertinent to young trees that have thin bark. Some sort of tree protection, such as a plastic tree guard or a section of wire mesh, is good insurance against such a problem. Such trunk protectors will also protect the trees from rabbits, voles and field mice during the winter. Be sure to place the base of the guard into the ground so that no bark is exposed.

Pruning

The truth is you do not have to prune an apple tree. An apple is well equipped to grow itself. Even if you do nothing you will get fruit. The practice of pruning focuses on improving the quality and size of the fruit and making the tree more consistently fruitful. Pruning can be a prickly subject. Many systems have been espoused and all have their adherents. The important thing to realize about pruning is that it is not so much sticking to one system but knowing techniques designed for different end results.

Pruning is a dwarfing process. In essence you are removing a portion of the tree's ability to produce food through photosynthesis. Once removed, the tree will strive to replace the wood and leaf area you have removed. The timing and type of pruning you engage in can have very different results. Pruning in spring encourages vegetative growth to replace the leaf surface lost. The more severe the pruning, the stronger the vigor of that growth. Pruning in midsummer tends to produce smaller lateral branching that is more directed toward flower bud production.

Simply put, pruning must be guided by a clear understanding of the need to balance vegetative growth with flower bud production.

Pruning Young Trees

When a tree is young, the first imperative should be to create a strong, spatially balanced framework of branches to support future fruit production. At the same time every grower wants their tree to come into production as soon as possible. To accomplish both tasks you

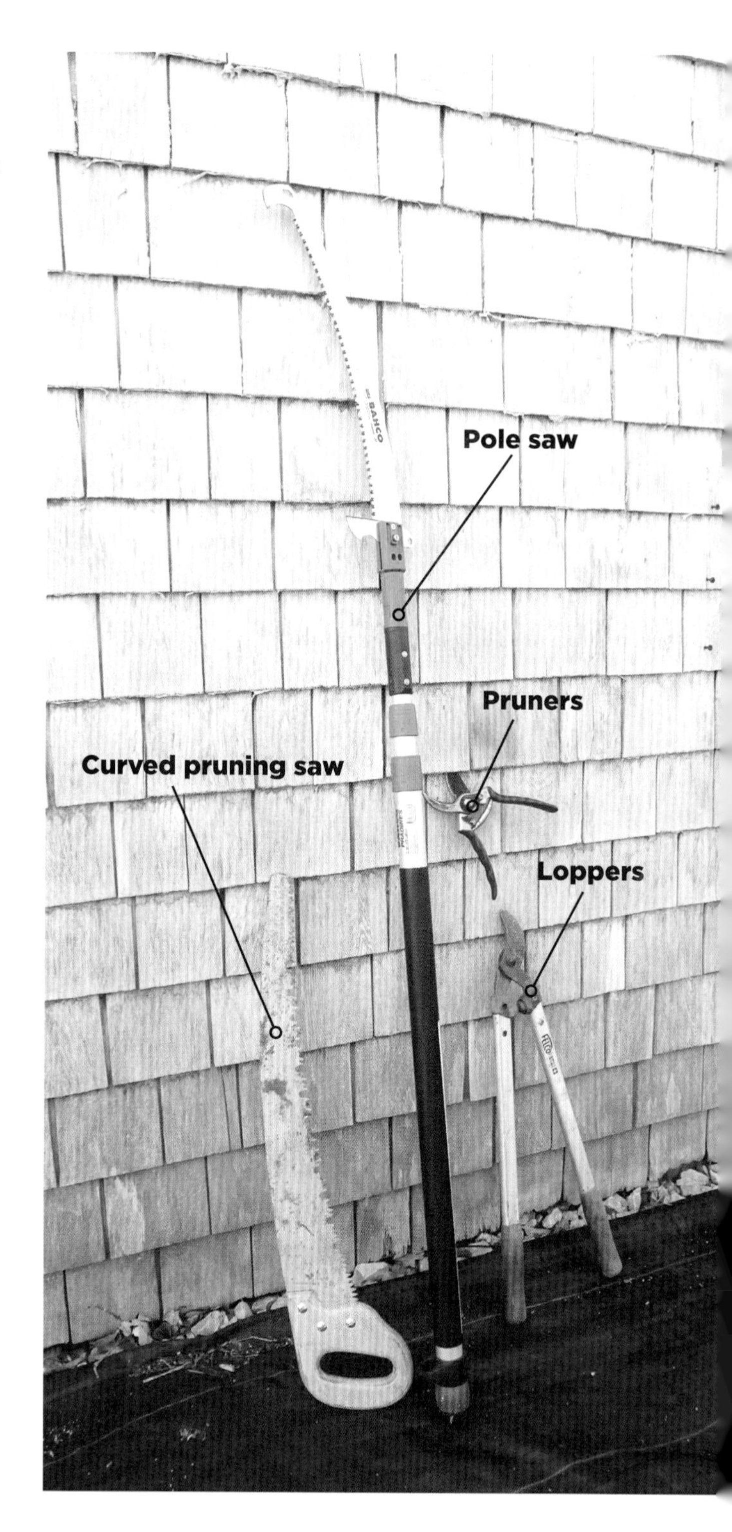

should prune only what is necessary in the early years. This can involve pruning some initial lateral branches to create good spacing between branches and to leave, as much as possible, an equal number of branches in each direction.

Branches that have bark inclusions in their crotches (where the trunk and branch meet) should be removed, as these can break under a load of fruit or snow and ice. Such breaks can tear a section of bark, which can lead to disease entry and a partial loss of connection between the tree's roots and its top. Any diseased or broken branches as well as branches that are crossing through the crown should also be removed.

Some growers prune back the tips of the main lateral branches to create stiffer branches, but caution should be used to limit the amount removed. Doing this in spring will create vigorous new growth and can delay fruiting. Pruning those tips in midsummer tends to produce smaller lateral branches and initiate fruiting spurs.

There is a great divergence of pruning techniques depending on what style of tree you are trying to create. In the past most growers placed their trees more or less equidistant to each other and open grown. Today most new commercial growers plant their trees in tightly spaced rows, with trees often only 3 to 4 ft. (1 to 1.25 m) apart, some even closer.

The most widely adapted system of pruning today is called the spindle tree. In this system the central leader is left untouched for the formative years, and the branches are selected to create a balanced pyramidal tree form, similar to the look of a classic Christmas tree but, of course, more open.

Strong tree crotch with no bark inclusion.

Weak tree crotch with bark inclusion.

High-density orchards use trees that have been budded or grafted onto dwarfing rootstocks, which will encourage earlier production and limit the size of the tree.

High-Density Pruning

Nearly all new large plantings of apple trees use a high-density system of management. The close spacing of trees and rows requires 1,000 to 1,500 trees per acre. This means initial capital costs for trees, posts and wires are high compared to lower-density orchards. The benefit is earlier and more affordable production, because all the trees are accessible from the ground or short ladders. Well-grown trees can produce large and colorful fruit.

Trees in high-density orchards are budded or grafted onto dwarfing rootstocks. These rootstocks force early production of fruit. Because much of the energy of the tree goes into forming seeds, there is a correspondingly lower production of vegetative growth. The training systems that have been developed are key to maintaining the size of closely planted trees and in keeping fruit production high.

High-density orchards most often use a support system of posts and wires. The trunk and the main upright stem (or central leader) of a new tree are staked vertically and allowed to grow upward until the central leader reaches the desired height, which on average is equivalent to 90 percent of the spacing between the rows. Optimally, this will take two to three years. When the desired height is reached, the top of the leader can be cut to a lateral bud. Some growers will bend the leader horizontally when it reaches the top wire of the support system.

As the tree ages and begins to produce fruit, one or two vigorous branches that have a diameter equal to or larger than half the diameter of the trunk are removed each year in favor of branches with smaller diameters. The cut used is called the Dutch cut or bevel cut. The branch is cut down and outward so that some of the base of the branch remains. This will encourage

◂ High-density plantings allow easy access to the entire tree — a boon for pruning and picking.

The Dutch cut leaves a basal area where regeneration of the cut branch can take place.

new growth to take the place of what has been removed. The remaining branches are trained to be from 50 to 90 degrees in relation to the trunk. This allows good light penetration into the tree and favors fruit production over vegetative production. When branches are horizontal or bent downward by the weight of the fruit, terminal growth is inhibited and fruit spur production behind the terminal is accelerated.

Consistent and logical pruning is essential to creating and maintaining a high-density planting. By keeping the trunk dominant to all lateral branches, the tree will remain narrow and productive.

As the tree ages there can be a preponderance of fruit spurs, and so some thinning of the spurs might be in order. This will cause fewer fruits to form and will tip the balance slightly toward the growth of new wood and new fruitful spurs.

Lower-Density Pruning

In an open-grown situation the central leader can be left growing to produce a taller tree. When the desired height is reached the terminal can be cut to a lateral bud or branch. Depending on the spacing of the trees in a row, the techniques used in high-density orchards can be used here to keep each tree relatively narrow, or the natural growth of the tree can be encouraged with attention paid to spacing the branches. As the tree ages, pruning is used to ensure light can penetrate the tree to promote earlier ripening and fruit color. At the same time it is important that the volume of the tree is filled with productive branches. An empty space cannot make apples.

Thinning

Although most modern apple trees bear fruit every year, apple trees in general tend toward biennial production, meaning that most cultivars will have a large crop one year followed by a small or nonexistent crop the next.

Some fruit may be lost in what is called the June drop. This is a form of self-regulation by the tree to ensure it can provide enough nutrients to the apples it keeps. However, even after the June drop, the tree still has enough apples that it is unlikely to overcome its biennial nature on its own. Thinning new fruit can create more balanced fruit production. Also, as the tree ages and more flower buds are produced on fruit spurs, the sheer number of apples will reduce the amount of water and nutrients available to each fruit. To create better fruit size and a more even production of apples year after year, the newly forming fruit must be thinned.

Large-scale producers rely on thinning sprays. A number of chemicals have been used in the past, but many are no longer used because they are caustic and can cause russeting of the fruit or are toxic to valuable insect predators. The insecticide Sevin® has been used for many years for fruit thinning. Mixing this product properly is critical to regulate the right amount of fruit drop. Newer formulations have also arisen, some that can be used in organic production. The most prominent today is Pro-Tone®, which contains s-abscisic acid (S-ABA), a naturally occurring molecule in plants that, for our purposes, causes the stomata on the undersides of the leaves to close. With reduced uptake of carbon dioxide, photosynthesis decreases and this results in what is called a carbon deficit. Since the young fruit are competing for the sugars produced by photosynthesis, the tree is triggered to drop a portion of its fruit. Attention to weather during the two to three days after application is important. Cloudy days and high temperatures will further decrease photosynthesis and will favor thinning. S-ABA is sprayed when the apples are 0.2 to 0.4 in. (5 to 10 mm) in diameter. A non-ionic surfactant is usually added to increase adhesion. Follow the mixing directions carefully so that you do not either underthin or overthin your trees.

Thinning can be done by hand if a small number of trees are involved. Each fruit spur can produce several apples, especially as the tree ages. As a rule of thumb, the fruit you

Removing the smaller fruits in a cluster will increase the size of the remaining fruit and will lessen the occurrence of insects such as leafrollers.

leave when thinning should be approximately 6 in. (15 cm) apart. This usually means leaving only one apple per spur. Each spur will have one larger fruit, called the king fruit. Unless damaged, the king fruit is the one you should retain and remove the others. You can use your fingers or pruning shears to do this, but be sure not to damage the spur itself.

The window for thinning is narrow. Once the apples grow larger than 0.8 in. (20 mm) thinning will be less effective and may not produce the desired outcome.

Espalier

Espalier is a specialized pruning method that creates what can be thought of as a two-dimensional tree. Espalier trees, usually grafted onto dwarfing rootstocks, are trained against a wall or fence. In most cases a one-year-old tree, often called a whip, is planted in spring, and the terminal of the central leader is removed. This results in several shoots developing under the cut. The most upright shoot is trained as the vertical trunk, and two others are trained as lateral branches, each in an opposing direction. This method is repeated each year until the desired form is achieved. The actual geometry of the tree can vary, with branches either at right or at lesser angles to the trunk. The tree can also be trained in the shape of a fan or candelabra. It is advisable to use a cultivar with a branching angle that best suits the desired shape.

The pruning methods involved in keeping an espalier tree productive but contained at the same time are similar to those used in high-density orchards. Vigorous wood is removed in favor of smaller-diameter branches that grow horizontally. A constant balance is maintained between encouraging vegetative growth, which produces new fruit spurs, and encouraging good production from existing spurs. It may be necessary to remove some older spurs as the tree ages. Thinning is also helpful to control production and encourage large, quality fruit.

To keep a balance between fruit and leaf, an espalier tree requires the grower to consistently remove vigorous wood in favor of smaller-diameter branches and thin spurs.

Pruning Older Trees

It is not uncommon for apple trees to grow 30 ft. (10 m) tall or more. If left alone, they can be a lovely addition to the landscape. Such trees can provide fruit as well, but if you are looking to produce larger, better-quality fruit, then the

Third-year Pruning

trees need to be thinned. As well, if you want to be able to reasonably work these trees, they will also need to be reduced in height. The idea of pruning older trees can be daunting, but it is certainly possible.

Reducing the height and thinning the branches of a large tree should be done over at least three years. Older trees recover more slowly from severe pruning. Just as surgery becomes more dangerous as a patient ages, so can drastic reduction put a tree's health at risk. You will be removing large portions of potential leaf surface, the solar collection organ of the tree, and that will compromise sugar production.

The first phase involves cutting back the trunk or trunks to the maximum height you wish to maintain. The trunk(s) should be cut back to an outward facing lateral branch at the desired height. If the cut is clean and close to the trunk you will not get too many new shoots, but there will undoubtedly be some. Try to remove these as soon as they begin growing. Not removing them will result in vigorous upright shoots called water sprouts. These upright branches will undo your goal of reducing the tree's height in short order.

The second phase involves removing vertical growing branches along the main laterals. These tend to produce vigorous vegetative growth. The branches growing horizontally or below can be left. These will be focused on producing fruit.

By the third year you can continue detail pruning to discourage upright growth, with the exception of areas that are open with no branches. Empty spaces do not produce fruit. Thinning should also be done to allow light into the center of the tree. (See diagram above.)

Within three years, with attention, you can have a smaller tree that is productive and has fruit that is harvestable using moderately sized ladders. Bringing a tree into such a condition is a satisfying journey. For a more complete guide to this process, called the Swiss renewal technique, try to find a copy of *Ecological Fruit Production in the North* (1983) by Bart Hall-Beyer and Jean Richard. This book is out of print but you might be able to find it on a secondhand bookseller's website.

6

Pests and Diseases

Dealing with Those That Want to Share in the Bounty

Nearly all plants contain the most important creation that ever evolved: chlorophyll. It is found within structures called chloroplasts. When struck by sunlight, chlorophyll has the ability to absorb parts of the light spectrum, particularly the red, orange, yellow and blue parts. It reflects the green part of the spectrum — the reason leaves appear green. The light provides energy to convert carbon dioxide from the air and water from the ground into sugars, with oxygen being the by-product of the reaction. Without chlorophyll we would not have food to eat or oxygen to breathe.

The animal and fungal worlds need nutrients, in particular sugars, to survive. There are myriad ways of obtaining food, but it always involves taking it from another source — another animal, another plant, another fungus.

Apples have a wealth of easily digestible nutrients, and they are high in sugars. This is why they are so valuable to us and also what makes them the inevitable targets of a large host of food-seeking life-forms. Everybody wants to eat. Dealing with these seekers is an important aspect of growing apples.

The production of organically grown apples is increasing slowly.

◂ Deer cause the most damage on young trees, whose new shoots they crave.

However, the lack of early research into organic methods has meant that organic growers have fewer tools at their disposal. Growing fruit organically is perhaps the most daunting challenge in horticulture. It requires an intimate knowledge of insect, fungal and bacterial species and their life cycles. It requires monitoring their populations and knowing how to time and apply the various methods of control. In short, it is not easy.

The good news is that we are entering a new era of pest control research. There are a number of new developments that offer a more environmentally friendly approach to control. This chapter covers old and new methods that can help keep the "spoilers" at bay.

Fruit Bud Stages

The life cycles of insects and diseases that infect apples are often referenced in relation to the stage of growth of the fruit buds. The geographic and even topographic location of the tree affects the amount of heat that accumulates and, thus, the arrival of diseases and insects. Seasons can also vary somewhat in heat accumulation, so the growth stage of the tree is a more precise way to determine when a particular insect or disease will be active than a specific time of year. Because many insects and diseases first appear before, during or just after flowering, the fruit bud stage is often used as a dependable measure of insect and disease development, and it helps the grower to know when a disease or insect is most vulnerable to protective measures.

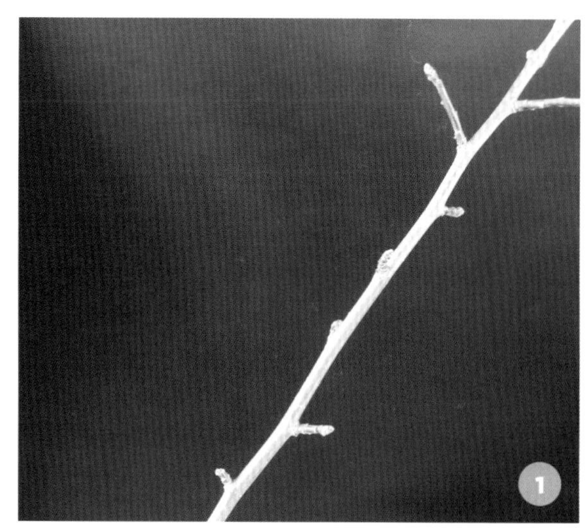

Dormant

Silver tip

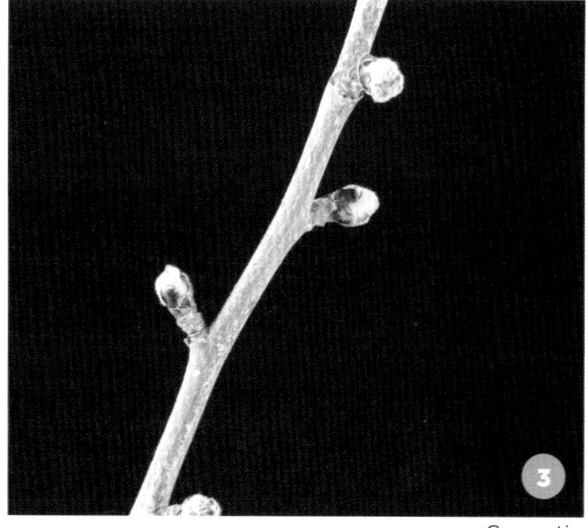

Green tip

Half-inch green

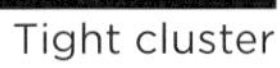
Tight cluster

First pink

Full pink

First bloom

Full bloom

Post bloom

Fruit set

Insects

Note: The products listed in the following section are made as suggestions and do not imply any warranty or guarantee of efficacy. When using any such product always refer to and follow the directions on the label.

Apple Maggot

Rhagoletis pomonella

The apple maggot is one of the major banes of apple growers across much of North America, particularly in the Northeast and Midwest. Apple maggot originally fed on bigfruit hawthorn, but when apples arrived in North America it quickly adapted to feeding on them. The adult is a fly, somewhat similar in appearance to, but smaller than, a house fly. On its wings is a pattern that many liken to a backward-facing F with a solid black line attached to the base of the F. This is a useful identifying mark.

Unfortunately this is one insect that you cannot use bud stage to predict. The adult fly emerges from the ground in the early summer (early to mid-July in Plant Hardiness Zone 4) during damp periods, usually in greatest numbers just after a heavy rain. They can continue to emerge until October. Some pupae may spend two to three years in the soil before emerging.

After emerging, the flies feed for seven to 10 days on aphid honeydew, bird droppings and various insect and plant exudates. During this period they become sexually mature and mate. Each fly can lay up to 500 eggs during its two- to four-week life span. The flies are attracted to the growing fruit and insert their eggs through the skin of developing apples. This entry point will later show up as a slight depression, like a dimple on the surface of the apple. After an egg hatches, the new larva begins burrowing through the fruit, leaving a narrow tunnel or trail that turns brown as it oxidizes. Larvae mature in 13 to 50 days, depending on outdoor temperatures.

An apple maggot fly.

Infected fruits most often drop early. After a few days on the ground the larvae exit the fruit and crawl below the soil surface where they pupate, remaining in the soil until the next summer. Most burrow no deeper than 2 in. (50 mm) from the surface.

To control apple maggot in commercial orchards, growers use a number of insecticides. The timing of sprays is often determined by the use of sticky traps. If the orchard is not already infested, the traps are placed on the perimeter of the orchard. They are often baited with ammonium acetate or ammonium carbonate.

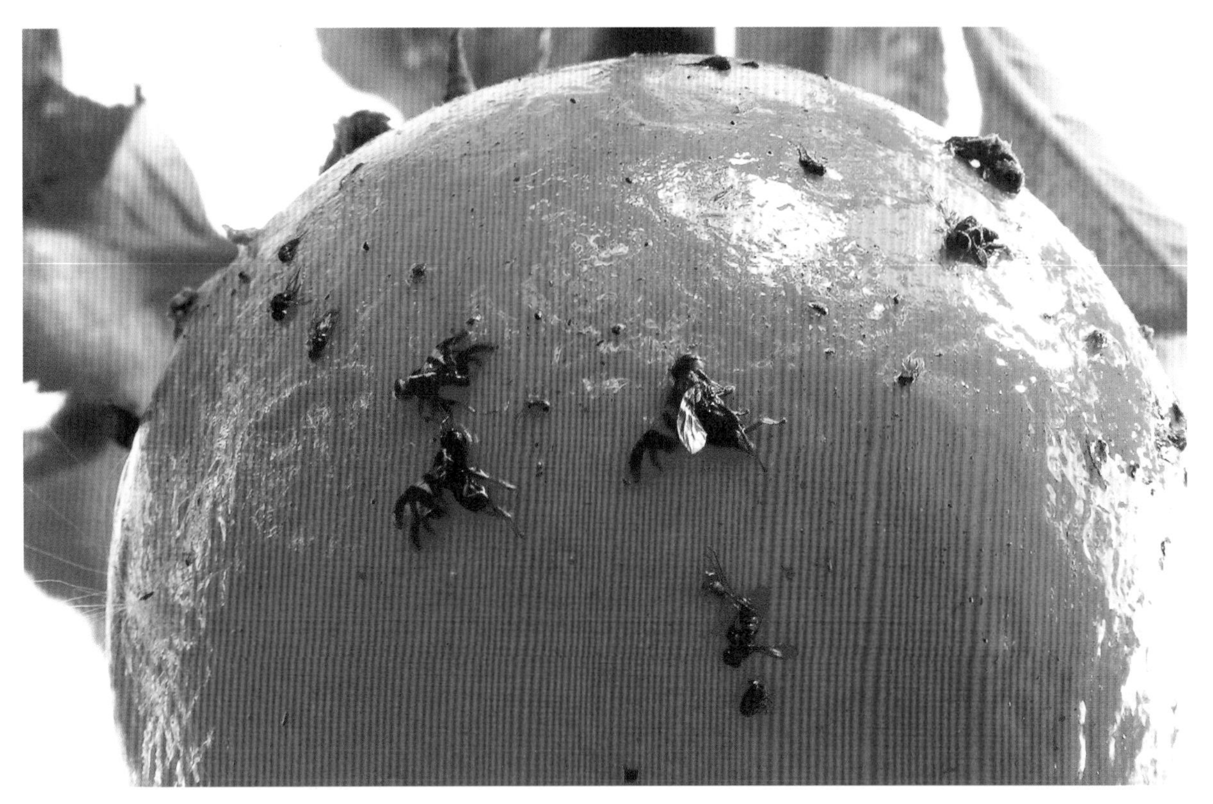

Yellow and red sticky balls can trap apple maggot flies and act as monitoring devices to time counteractive measures.

Traps will work without either, but they will be more effective using one of these baits. Research has shown that ammonium acetate is the more effective of the two.

A teaspoon of bait is placed in a small plastic vial or film canister with holes in the top that allow the chemical volatiles to escape. The vial is hung below the trap, and the bait should be replaced every seven to 10 days. Flies do not lay eggs until eight to 10 days after emergence. The first spray is recommended seven days after the first fly is captured. As flies continue appearing until late in the season, monitoring and control must be maintained.

There are several methods that help in control and do not involve insecticides. A popular method that is particularly useful for the home orchard uses plastic or wooden balls, most often red or yellow. Yellow balls (or yellow sticky cards) are more effective at trapping immature (unmated) flies as they emerge from the ground. Red balls are more effective at trapping adult flies. The balls are coated with a sticky substance, such as Tanglefoot®, and can be baited. Once the fly lands on the surface it is caught and cannot lay eggs. Hang several sticky traps in the orchard to act as monitors. Once a fly is trapped the rest of the sticky balls can be hung.

Because the traps become covered with all manner of insects, some growers cover the traps with a clear plastic "sock" and apply the Tanglefoot® to the plastic, changing it when required, usually after two to three weeks.

The balls should be placed at approximately eye level, just inside the outer layer of the canopy. It is recommended you remove leaves in the immediate vicinity of the balls to better expose the traps and prevent leaves from becoming stuck to them. The size of the tree will determine the number of balls. A small dwarf tree may only require one or two traps, a semi-dwarf tree three or more traps and a large tree six to eight. Research suggests one trap per 100 apples is optimum.

Physical barriers placed over the fruit are also effective for protection; however, if your tree or orchard is bigger this may not be practical. Barriers placed over the developing fruit can be clear plastic bags (note that humidity can develop inside), paper bags (these will need to be replaced after rains and apples will not redden) or disposable nylon socks. Be sure that there is no entryway into the apple.

Several newer products are also available. Surround® is a kaolin clay product that can be sprayed on the apples and foliage. The fine layer of clay particles interferes with egg laying (oviposition), deters insects physically and may make the fruit unrecognizable to pests. Interestingly, the clay does not seem to interfere with photosynthesis. It is inert and nontoxic, though it can deter beneficial insects as well. This product needs to be applied several times throughout the season.

Spinosad is a newer product that has been certified for organic fruit production. It is derived from two complex organic compounds produced by certain soil microbes. When ingested, these compounds are toxic to insects. (It is not considered harmful to mammals, birds or aquatic life.) Because it is only toxic if ingested, not simply by contact (as many insecticides work), it is thought to be less harmful to beneficial insects that are not consuming the leaf or fruit tissues. Spinosad is the active ingredient in products such as Entrust Naturalyte™, which is registered for use on fruits. Spinosad can be used to control many insects such as leafroller, codling moth, fruit worm and apple maggot. For apple maggot, controlled timing, as described earlier, is critical. It should be sprayed when the first flies are trapped. Entrust Naturalyte™ (also called GF-120 commercially) can be applied with a backpack sprayer with the spinner in the nozzle removed so that it is delivered in large droplets. Whip the wand back and forth a few times so that you hit the undersides of the leaves. This helps to keep light rain from washing it away. Only a small amount is needed, and in larger orchards spraying every fourth tree or so is sufficient to attract the flies. Once they land they eat the product and die. Spraying should be repeated after heavy rains or approximately once a week until nearly harvest. We have found this method extremely effective.

Parasitic nematodes are microscopic, thin, worm-like creatures that lay their eggs in the pupating larvae of many insects, of which the apple maggot is one. The host dies, and thousands of tiny new nematodes are released that then go in search of new prey. Several species of nematodes in the genus *Steinernema* are effective and have been commercialized. The nematodes are shipped directly to the user. The medium containing the nematodes is mixed with water and injected into the soil under the trees. The nematodes should be used quickly after arrival to ensure they do not die and should never be exposed to direct sunlight. They can be injected into the soil any time after the ground warms. Though perhaps not feasible for larger orchards, the home gardener might find this method effective.

Codling Moth

Cydia pomonella

The iconic picture of a worm poking its head out of an apple is a representation of the codling moth larva. This worm can do catastrophic damage to the apple, rendering it useless for anything but chicken feed — though chickens do love worms!

The codling moth larva burrows to the center of the fruit to feed.

Codling moths overwinter in cocoons spun in deep crevices in the bark of trees, on wooden bins, pallets or similar objects located near orchards, and sometimes in the ground. The larvae pupate just before the buds show first pink, and the adult moths emerge around full bloom. They continue to emerge for two to six weeks, depending on the temperature. Some adults emerge later, in late July or August if conditions are right.

The adult moth that emerges from the cocoon will mate as soon as temperatures at twilight are above 60°F (15°C). The moths are 0.4 to 0.5 in. (10 to 12 mm) in size, brownish gray and have a characteristic copper-colored spot on the end of each forewing. Eggs are laid on the surface of the apples or on adjacent leaves. When an egg hatches the larva spends some time wandering the fruit surface until it finds the spot where it wants to bore through the skin. This entry hole can be recognized by the brownish grainy residue in the hole left by feeding. The larva moves toward the core of the apple and enlarges as it feeds, growing up to 0.8 in. (20 mm) in length. The larvae are white, sometimes pinkish, with a black or mottled-black head. After reaching maturity, the larva leaves the fruit and wanders to find a suitable place on the tree or nearby to spin a cocoon.

Various methods can help reduce the population of this pest. Sanitation in the orchard can be a key component of nonchemical control. The entry hole of the codling moth is blocked by the brownish-red leavings of digested apple. Once recognized, this telltale marker makes it

Although similar in appearance to damage caused by the European sawfly, codling moth damage shows up later, and the frass does not smell.

easy to walk through the orchard and remove the affected apples. If done thoroughly, their removal and destruction will greatly decrease the following year's population. Burial, complete crushing or proper composting will destroy the larvae. Tree trunks and main branches should be inspected in winter for cocoons. Wooden materials should also be kept out of the orchard area.

To monitor and trap late-emerging larvae, corrugated cardboard can be wrapped around trunks and main branches. The larvae are attracted to these "hiding places." The cardboard can then be removed and any cocoons destroyed. These should be checked and changed several times during the season.

Codling moth traps with lures are used in orchards to monitor the arrival of moths. These are also available to the home gardener. The traps lure male moths by releasing a mimic of the female's attractant pheromone. Traps should be set out just as flower bud movement is detected (silver tip).

Isomate®-C Plus is a mating disrupter that can be applied to large orchards. After petal drop, dispensers are placed in the upper canopy of the tree in shaded sites. This material is not recommended for orchards smaller than 5 acres (2 ha).

Virosoft™ CP4 is a biological spray that contains a virus that gradually destroys the larvae. It is applied when the eggs are hatching. It must be ingested to work, so complete coverage is essential. Because it takes time to work there will be some damage to fruit before the larvae die. This is another control that is suitable only for larger orchards.

Plum Curculio

Conotrachelus nenuphar

The adult plum curculio is a beetle with a curved snout, a member of the weevil or snout beetle group. It is 0.15 to 0.23 in. (4 to 6 mm) long with a snout that is approximately a quarter of its length. The curculio varies in color from brown to gray with gray and white patches. It has four bumps on its back. The larva is white, legless and up to 0.4 in. (10 mm) long with a brown head.

This insect attacks all the pome fruits (apples, pears, plums, etc.) as well as soft fruits, including gooseberries and chokecherries. It overwinters in ground litter, woodpiles and similar places. When apple blossoms reach first

Curculio damage results from egg laying. The scar starts out fan shaped but can enlarge as the apple matures.

pink through petal drop, the adult flies to the apple trees and begins feeding on leaves, buds and eventually young fruits. It feeds for four weeks. The beetle will create a crescent-shaped opening in the skin of the apple and insert a single egg. The egg hatches in approximately one week, and the larva feeds inside the fruit. After two weeks the larva drops to the ground and pupates. New adults emerge in two to three weeks and begin feeding on fruits until fall, after which they move to wintering sites.

Damage done by the plum curculio includes the crescent-shaped scar on fruit created by the egg-laying activity, as well as some notching on petals and leaves. The adult plum curculio makes cavities in the fruit it feeds on that can decay by harvest.

This insect can be difficult to control with organic methods. An older, and still effective, method is to spread a sheet under the tree just before sunrise and shake the tree. The beetles will drop onto the sheet where they can be captured and killed. This should be done between first pink and petal drop.

Collecting any dropped fruit will help break the cycle by capturing the larvae before they have a chance to pupate.

Although this insect can be a problem, we can be thankful that the apple is not preferred by this beetle and that, at least in the north, there is only one generation per year.

Tarnished Plant Bug

Lygus lineolaris

The tarnished plant bug winters under bark, leaves and debris. The adult emerges early in spring and feeds on the tree's flowers and developing fruit. The insect is 0.23 in. (6 mm) in size, oval-shaped and varies in color from green to brown. It also has a characteristic light-colored Y on its back. It can be mistaken for a wasp. The

A tarnished plant bug.

The black dimples or pits are caused by the tarnished plant bug's feeding.

nymphs that hatch from the eggs are 0.04 in. (1 mm) in size. They molt (cast off old skins for new growth) five times before becoming adults. Every life stage of the bug feeds on apples, and as there can be several generations during the growing season, all stages may be in the orchard at any one time.

These bugs are extremely mobile and will fly off with the least bit of disturbance. Their saliva contains toxins that tend to distort growth, causing premature dropping of flowers, distorted terminal growth and pitting of fruit. Their feeding can also create a partially peeled look on the surface of the apple called catfacing.

While controlling the insect can be difficult, some advocate removing or cutting weeds around the orchard to reduce wintering sites. Others suggest planting alfalfa near the orchard to lure them away from crops as they prefer alfalfa to most plants. There is also a parasitic wasp called the European nymphal parasitoid (*Peristenus digoneutis*), which lays an egg inside the tarnished plant bug nymph (immature stage). These wasps can be purchased from insectaries.

Speckled Green Fruitworm Moth
Orthosia hibisci

Although there are several species of butterflies and moths (Lepidoptera) that are grouped under the term "fruitworm," the most prevalent is the speckled green fruitworm moth. The adult moths are 0.6 in. (16 mm) in length with grayish-pink forewings. Each forewing has two purplish-gray spots. The underwings are slightly lighter in color. The adults fly at night, appearing when the fruit buds are at green tip stage up to the pink stage.

The moths lay their eggs on twigs and newly emerging leaves. They can lay several hundred eggs but generally lay only one to two per site. When the larvae hatch they are greenish in color. They pass through six instars (molts), eventually growing to 1.2 to 1.6 in. (30 to 40 mm). They have green bodies and heads with several white lines running the length of their bodies and two more prominent lines on the sides. White speckles are distributed in the green areas between the lines. They begin

A speckled green fruitworm larva.

feeding on new leaves and fruit buds and are often found feeding inside leaves they have pulled together with webbing.

One to two weeks after petal drop, the larvae drop to the ground and burrow 2 to 4 in. (50 to 100 mm) below the surface, where they pupate.

Most fruit buds that are attacked drop prematurely, but those that continue developing into fruit will show deep corky scars and indentations that are virtually indistinguishable from the damage done by the oblique-banded leafroller.

Most commercial orchardists spray organophosphate pesticides between the green tip and pink stages of fruit bud development and then again after petal drop. Organic options include *Bacillus thuringiensis var. Kurstaki* (Btk), which is sprayed during bloom and/or at the petal drop stage, or Spinosad, which is applied at first pink or petal drop. Btk is a bacterium that infects the gut of the larva, so it must be ingested, meaning coverage should be complete and applied when rain is unlikely in the near future. Spinosad (an ingredient in Entrust Naturalyte™) should be used alternately with Btk to avoid any development of resistance to the products.

Obliquebanded Leafroller

Choristoneura rosaceana

The eggs of this pest winter in bark crevices and branch crotches. When temperatures reach 50°F (10°C) they begin hatching, about the time flower buds are in the tight cluster stage. The larvae move to terminal growth and fruit spurs to feed on foliage and emerging flowers. After feeding for about a month the larvae attach webbing to the leaves and roll the foliage for protection against predators and to form a loose cocoon. This also protects them from sprays.

Leafrollers protect themselves from predators such as birds by rolling up leaves with webbing.

After four to five weeks the adults emerge to mate and lay eggs. Although most eggs are laid in the upper canopy, some are laid in the lower areas of the trees. Once hatched, the new generation feeds on terminal growth but also the new fruit, particularly where fruit is in clusters or where leaves offer cover. The larvae spin webbing that attaches the leaves to the surface of the fruit. The larvae also produce silken threads that allow them to float in the wind to other trees. Feeding will produce a range of damage to the fruit surface, from small "pinpricks" to larger corky scars.

In some commercial fruit growing areas leafrollers have become resistant to many of the older organophosphate insecticides and even to pyrethroids. Applying dormant oil to the trees before the silver tip stage can help reduce wintering populations. Because they are the larvae of a moth, they can be controlled by spraying the tree with *Bacillus thuringiensis var. Kurstaki* (Btk) at petal drop.

Thinning fruit clusters to a single fruit can help prevent damage to fruit. There are also many parasitic wasps that attack leafrollers.

Redbanded Leafroller

Argyrotaenia velutinana

The yearly cycle of the redbanded leafroller begins with the emergence from the ground of small moths with a characteristic red band on their forewings. These weak flyers usually lay egg masses on trunks or low scaffold branches. These masses contain approximately 40 eggs

A redbanded leafroller larva.

each and resemble yellow or cream-colored overlapping scales. The larvae hatch just after petal drop and begin climbing into the tree, usually attacking water sprouts first.

They skeletonize leaves from the underside, mostly around the midrib vein, producing a small web around them for protection. As they grow older they begin feeding on leaves that they have folded and attached to the surface of developing apples. The surface of the fruit is eaten as well. As fall approaches they fold leaves around themselves and pupate, eventually dropping to the ground to await spring.

Damage on developing fruit can be deep,

The redbanded leafroller's damage is often found on the side of the fruit. The insect will attach a leaf to the developing fruit for protection and eat the surface of the apple.

causing corky scars and deformation. The later brood that feeds on older fruit usually causes shallower scars.

The redbanded leafroller has many natural enemies. It is interesting to note that this insect was not considered a major pest until the introduction of DDT into commercial orchards killed its natural predators. Pheromone traps are used to measure infestations and could be used to help control high quantities. Banding trunks and branches above the egg-laying height with Tanglefoot® might be effective if placed just before petal drop.

The most effective prevention method is to spray *Bacillus thuringiensis var. Kurstaki* (Btk) directly after petal drop.

Apple Red Bug

Lygidea mendax

This insect spends the winter in bark crevices and hatches out just before blossoming. The insect is red with a brown midsection and black legs. It is similar in many ways to the tarnished plant bug but has only one generation per year.

Although superficial, the damage by the apple red bug spoils the aesthetics of the fruit.

The apple red bug feeds on foliage, which becomes distorted, as well as developing young fruit. The damage is seen at harvest as small irregular patches of russet.

This insect does little damage to foliage and only marginal damage to the fruit. It can be controlled with a dormant oil spray just before the pink tip stage. The refined oil smothers the eggs.

European Apple Sawfly
Hoplocampa testudinea

This insect winters in a cocoon in the soil just under the surface. The adult emerges between the green tip and petal drop stages. The adult fly resembles a wasp. The upper body is dark and shiny, and the lower body, legs and head are orange. The wings are translucent with black veins.

The adult lays its eggs on the petals. Once hatched, the larvae go through five instars (molts) until they are 0.3 to 0.43 in. (8 to 11 mm) in length and pale brown with a black head.

Sawfly damage from early feeding on the young fruit shows up as curving scars that emanate from the basin of the apple.

The larvae tunnel into and through the developing fruit. The first instar produces a scar that spirals from the basin (the end opposite the stem end) of the fruit. The second instar

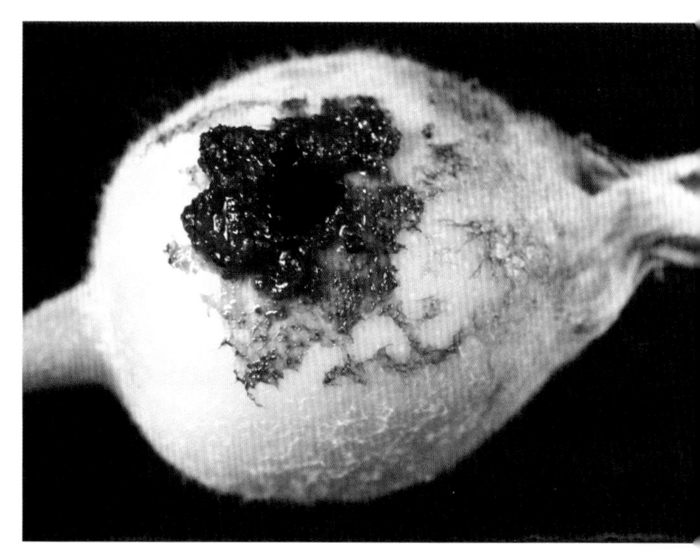

The exit hole of the European sawfly with its strong-smelling frass. The affected fruits usually drop by the end of June in the north.

makes an exit hole. Frass, which looks like red-brown sawdust, becomes visible. Later instars enlarge the hole and produce more frass. The larva moves toward the central seed cavity. Most affected fruit will drop in June.

The damage is similar to that of the codling moth but shows up two to three weeks after petal fall, whereas codling moth damage does not show until five weeks or later after petal fall. The codling moth larva is also larger and pinkish white in color. Sawfly frass is strong smelling, whereas codling moth frass is not.

White sticky cards are used to trap and monitor sawflies from the tight cluster stage to petal drop stage. The white mimics apple blossoms. Traps should be put on the south side of the tree at eye level.

Sprays are usually applied at first pink and shortly after petal drop. After petal drop is considered the ideal time to apply pesticides. Organic formulas include rotenone, ryania or Spinosad. Though this insect has no known predators in North America, there have been controlled introductions of *Lathrolestes ensator*, a species-specific larval parasite, into orchards in Ontario and Quebec. It is hoped, once established, the parasite can be introduced into other areas.

Buffalo Treehopper

Stictocephala bisonia

This insect is named for its vague resemblance to a bison (often called buffalo). It is light green in color, and the sharp hood over its head makes it resemble a thorn or protuberance on the stems, helping to camouflage it. Adults emerge in the fall and make slits in the bark of new branches to lay eggs. The emerging young feed on the sap of trees. In large numbers they can cause stunting to the tree, particularly a young tree, and give a rough appearance to the bark.

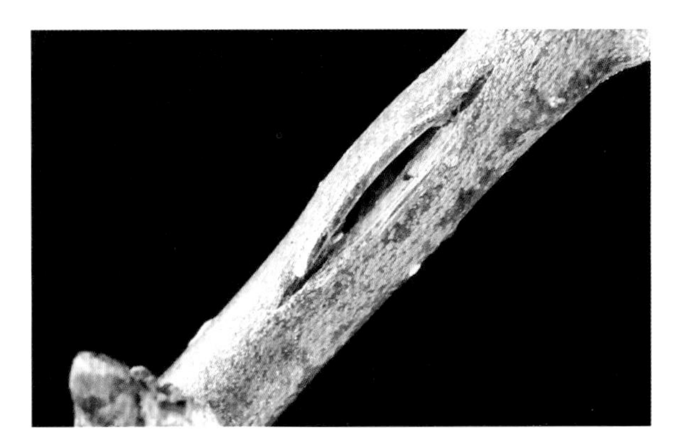

The buffalo treehopper slices open the bark of young shoots to lay its eggs. Although not usually a big issue, a large infestation can severely affect a shoot's growth.

This is a minor pest but can be more significant if the orchard has legumes such as alfalfa growing in or near it. The buffalo treehopper prefers alfalfa to other species. If your area is infected with this insect, it is advised not to plant legumes in or near the orchard. You will notice here I am contradicting advice I gave for introducing nitrogen into the soil and reducing the numbers of tarnished plant bugs. Alas, nothing is simple!

Aphids

Dysaphis plantaginea, Aphis pomi, Eriosoma lanigerum and *Rhopalosiphum fitchii*

Several species of aphid feed on the tender new growth. They include the rosy apple aphid (*Dysaphis plantaginea*), the apple aphid (*Aphis pomi*), the woolly apple aphid (*Eriosoma lanigerum*) and the grain apple aphid (*Rhopalosiphum fitchii*).

These small soft-bodied insects are flightless most of the year. They suck sap from rapidly growing tissues, usually the terminals. When an aphid penetrates the veins of the plant, the internal pressure of the plant pushes sap through the aphid and out structures called cornicles or tubercles, at the rear of its abdomen. This sugar-rich material, called honeydew, becomes blackish as molds grow on it and often attracts ants, who feed on the honeydew and protect the aphids against predators.

In great numbers aphids can reduce the vigor of young trees.

In organically managed orchards several

Aphids on the underside of a leaf.

The curled appearance of these leaves indicates an aphid infestation. This colony has been destroyed by predators.

species keep aphids under control. Lady bug larvae, syrphid fly larvae and lacewing are common predators. The chalcid and braconid wasps also attack aphids. Using pesticides to control aphids often reduces the populations of these predators and parasitic wasps, too, and because the predators reproduce more slowly, this can result in a strong resurgence of the aphid population. As well, aphids can develop resistance to pesticides over time.

The best management practices include avoiding excess nitrogen in the soil, as it tends to create rapid succulent growth, which attracts aphids. Water sprouts, because of their vigor, tend to attract aphids, so removing them early will aid in preventing aphids. Syrphid fly populations can be increased by growing buckwheat in or near the orchard. Several of these predators can be purchased from insectaries as well.

Scale Insects

Quadraspidiotus perniciosus, Quadraspidiotus ostreaeformis and *Lepidosaphes ulmi*

There are three species of scale insects commonly found in orchards. They include San Jose scale (*Quadraspidiotus perniciosus*), European fruit scale (*Quadraspidiotus ostreaeformis*) and oystershell scale (*Lepidosaphes ulmi*).

Fertilized females winter on young twigs. They begin feeding when sap rises in the tree. The females give live birth to young called crawlers, which leave the protective wax coating that covers the female and move to new feeding sites, usually within 3.3 ft. (1 m) of the

Scale insects on a branch.

female. Once they insert their feeding tubes through the bark and start feeding, they begin creating a wax covering that protects them from predators and desiccation. This covering makes them difficult to control with chemicals.

Males arrive in summer to mate with the new scale insects. Once mating occurs the males die.

Severe infestations of scales can kill limbs and damage fruit. Most damage shows up at the calyx or stem end of the apple. The gray scales are usually surrounded by a reddish or purplish discoloration.

Scale insects are most vulnerable early in spring, just before the tree's growth begins. A spray of dormant oil will kill many scale insects. When crawlers leave the female scale insects, they are vulnerable to sprays as well. Monitoring for crawler movement is often done using black electrical tape placed sticky side up around young stems. The tiny yellow or white crawlers are caught and show up against the black tape.

Natural predators, such as birds, usually keep scales in check. Spraying with insecticides can reduce populations of predators and cause outbreaks of scale insects. If you have only a few trees and they are accessible from the ground or with a stepladder, you can physically remove the scales by gently rubbing them off using a nylon scrubber you might use to wash pots.

Fall Webworm Moth

Hyphantria cunea

Fall webworms appear in late summer or fall. When eggs hatch the young caterpillars create a web toward the ends of branches that encloses the foliage. They consume the leaves within the web and then move outside to feed on nearby leaves. These nests should be removed by hand and destroyed when first noticed. Their numbers vary from year to year, and if caught early they do little damage.

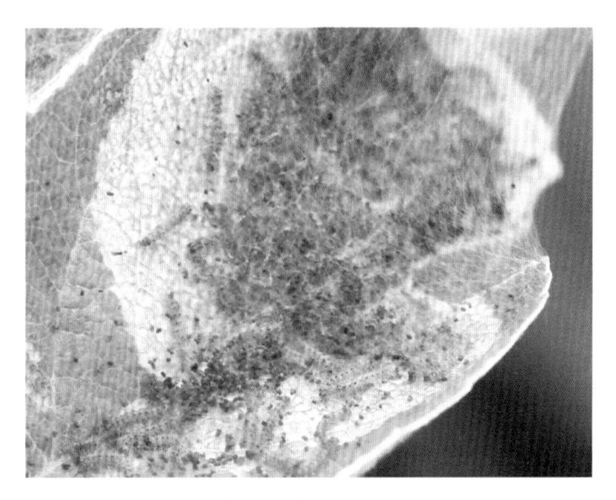

Fall webworm caterpillars inside a web.

Eastern Tent Caterpillar

Malacosoma americanum

The moth stage of this insect lays its eggs on the side of the tree that faces the morning sun in early spring. The eggs appear like tiny barrels stuck together, and masses of eggs partially surround small twigs. The entire cluster is coated with a substance called spumaline. This gives the whole egg mass a shiny appearance and offers some protection from marauding wasps and other species that feed on the eggs. The spumaline, however, offers no protection from a thumbnail pushed against the mass. It will pop off easily, and once on the ground it will be eaten by something.

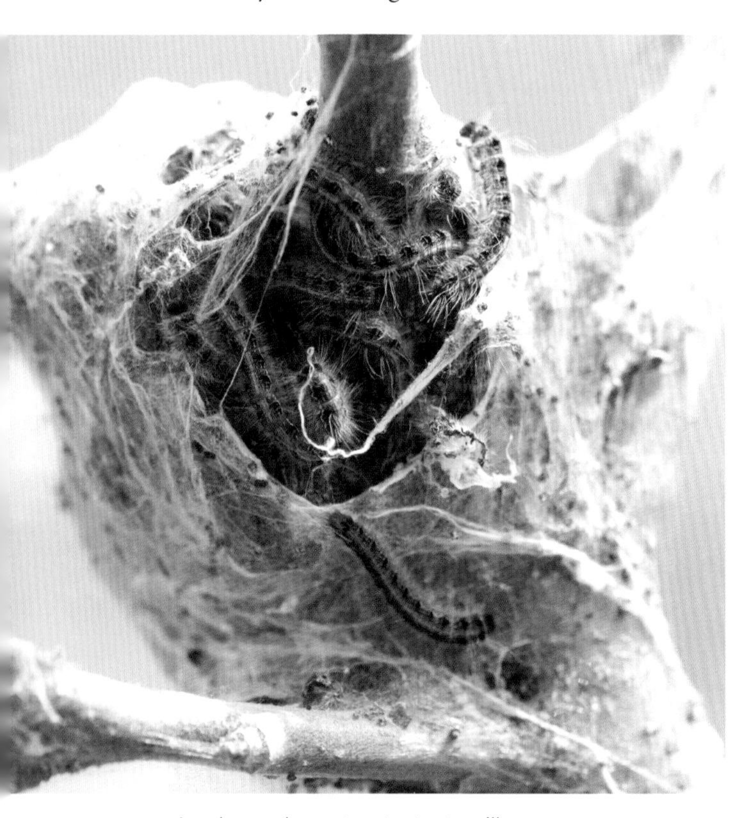

A web spun by eastern tent caterpillars.

The young caterpillars spin a web in the crotches of branches. If caught when first noticed and destroyed they will cause little damage. If allowed to feed without interference they can defoliate large sections of the tree. The nests are made of layers that trap the warmth of the morning sun. The young caterpillars can be seen in a group toward the center, where temperatures rise to high levels, which allow them to digest the young leaves that are its food.

Keeping a sharp eye out for the egg masses and removing the tents by hand as soon as they are noticed is the easiest and best solution. Early mornings or late evenings are the best time to catch them all within the nest. Throw the nest to the ground and grind it with your heel. The pest becomes fertilizer. Wear gloves and heavy boots if you are squeamish.

Roundheaded Appletree Borer

Saperda candida

This insect is a killer of trees, specifically young trees. In northern areas the beetle emerges in early June and continues laying eggs until July or even August. It has long antennae and long, rectangular hard forewings that have alternating tan and white stripes running down its back. It creates a small slit in the bark of young trees near ground level, into which it deposits its eggs. More than one beetle may visit the same tree.

As the larva develops, it begins feeding on the cambium layer just beneath the bark. The feeding disrupts the flow of nutrients to the root system and weakens the trunk. You can often

An appletree borer larva. Appletree borers are one of the most destructive pests.

see orangish-brown frass (its sawdust-like excretions) on the outside of the tunnels. The tunnels can go in any direction and sometimes will connect with other tunnels from other larvae.

The larvae will overwinter in the trunk and begin feeding again the next spring. If not caught in time, the tree will most often die. You might see a prodigious flowering in spring, but the tree will collapse before the fruit develops. Other symptoms are sunken and discolored bark, a generally poor appearance and smaller leaves.

The key to preventing damage is continual inspection of young trees at their base and even into the ground a short way. You must find the slits where the larvae are feeding and kill them. This can mean delving down their tunnels into the tree with a sharp pointed object. Michael Phillips, an organic grower from Maine who has written several excellent books on organic fruit production, prefers a 0.37-in. (9.5-mm) spade bit as his hand tool of choice. It has the perfect shape for the job.

Carefully dig into the tunnels until the grub is found and kill it. If not, its numbers will increase and do more damage to more trees in the future. It also attacks wild trees that are related, such as hawthorn and mountain ash. If these grow near the orchard it is wise to remove them.

A good prevention method is to coat the bottoms of young trees with latex paint. Do not use enamel or acrylic as it may damage the tissues. The paint helps deter egg laying. Others have also suggested tying plastic or aluminum screening tightly around the trunk. Be sure to bury it slightly into the ground. Another good practice is to remove tree guards during the growing season to weed around the base of the tree and inspect the trunk. Even with these measures you must check your trees frequently.

This is one insect that should be shown no quarter. Eternal vigilance is required.

Flatheaded Appletree Borer
Chrysobothris femorata

This pest attacks many species of deciduous trees and shrubs, including beech, maple, oak, dogwood, cotoneaster and willow. It can be a serious pest of apples, in particular young trees.

The adult emerges from late spring into midsummer. It is a bronzy-green beetle with two distinctive lighter wavy lines on its wings. The first section of the thorax, not actually the head, is wide, thus giving the insect its name. The beetle lays its eggs under sections of bark or in

Appletree borers must be destroyed as soon as their entrance holes are discovered.

crevices. The larvae hatch out and bore under the bark and feed on the cambium. The larvae winter in the feeding chambers. The following spring the mature larvae bore into the heartwood to pupate, emerging to begin the cycle again.

In older trees this can cause the tissues above the feeding chambers to die, and branches above these sites are often injured or killed. In young trees the borers often girdle the entire trunk, and within a year the tree collapses and dies.

Inspecting the trunk near the ground in late spring and early summer can help detect the presence of the adult beetle. If the presence of sawdust is noticed and a hole detected, thin wires can be inserted to skewer the larva or bleach can be squirted into the hole with an eye dropper.

Keep your trees in good health to create good vigor and maintain a close eye on trees during the spring and summer. Vigilance and early detection are the keys to preventing this destructive insect. Twelve species of parasitic wasp attack this insect, and woodpeckers eat many larvae, though they might make a mess of your tree as well.

Fungal Diseases

Apple Scab

Venturia inaequalis

It is safe to say that apple scab is one of the most challenging diseases to overcome in apple production. It affects most apple cultivars to varying degrees in all but the most arid growing areas.

Apple scab overwinters on fallen leaves and fruit. In early spring, fruiting bodies develop that release spores, called ascospores, as the tree comes out of dormancy, peaking at the flower's first pink stage. The greatest percentage of spores is released in the early morning after a rain has wetted the spore bodies.

The spores land on developing leaves and fruit, and successful spores form germ tubes that penetrate the surfaces. A web of mycelial threads grows between the cuticle and epidermal cells. These threads puncture the cell walls and feed on the sugar-rich sap inside. Most initial infection occurs on the undersides of the leaves. The early appearance of these sites will look olive green to brown, gradually turning deep brown to black as the infection site ages.

Apple scab can occur on both leaves and fruit. Lime sulfur remains the fungicide of choice for organic orchards.

Apple scab may be only skin deep, but it renders the fruit unpalatable, and affected apples will not keep well in storage.

At this point the fungal mycelia penetrate the cuticle and release asexual spores called conidia. The moisture of morning dew and rain showers helps spread the spores, which create secondary infections. This process continues until leaf fall.

The black to brown lesions on leaves can cause premature leaf drop during the summer. This reduces food production and weakens the tree. Lesions on a fruit render it virtually unmarketable and will cause early breakdown in storage.

Scab will infect at temperatures between 41°F and 77°F (5°C and 25°C), when leaves and fruit have been wet for at least eight hours. The longer the trees stay wet the higher the infection rate. Ideal conditions for scab are when temperatures are around 68°F (20°C) and leaves have been wet for more than 18 hours.

The easiest solution for new orchards is to choose scab-resistant cultivars. This resistance can range from partial to near absolute. Many new apples that have emerged from breeding programs are of excellent quality and virtually immune to scab. There are many others that have good resistance to scab. Such apples will have a good percentage of usable fruit at harvest.

If you want to grow those that are susceptible to scab, and there are many, you will need to instigate certain procedures to reduce its impact.

Because scab overwinters in the leaf litter, it helps to remove and destroy the leaf litter or alternatively to provide a deep mulch cover that buries the litter, thus not allowing the spores to reach the air. This, however, does not eliminate spores that arrive from other sources, but it can significantly reduce the number of spores within the orchard.

Growers in the past have relied on fungicides such as Captan or Benlate® (benomyl) to kill spores on the surfaces of leaves and fruits. These are being replaced with new materials such as sterol inhibitors and strobilurins, but increasing resistance is rendering them less effective.

The organic grower has only a few materials that are suitable and effective. Older materials such as Bordeaux mix or copper can be effective, but they often cause burning or russeting of the fruit and leaves.

Bordeaux mix is a blend of copper sulfate, hydrated lime and water. The copper has fungicidal properties and the hydrated lime "burns" the spores. Copper, usually applied in the form of copper sulfate, is a fungicide. Using these materials runs the risk of creating unacceptable levels of copper in the soil, which can be detrimental to soil organisms.

Lime sulfur is a blend of elemental sulfur and hydrated lime. The lime makes this a relatively caustic material, though it does enable the grower to treat infections that have already occurred. It can produce burning and russeting of leaves and fruit.

At present most organic growers rely on sulfur as their main deterrent. There are two forms in use. Wettable sulfur is a combination of elemental sulfur and bentonite clay. This material is difficult to apply as it must be agitated constantly in the sprayer or it will clog the spray orifices. When it lands on the leaf and fruit surface, the oxidation compounds create an acidic layer that inhibits germination of the spores. As with most other fungicides, it must be reapplied after rains to maintain continuous protection. It should not be used within a month of any dormant oil applications.

The second form is called flowable sulfur. This material is an emulsion, not a suspension like wettable sulfur. Flowable sulfur is easier to use, distributes itself over the surfaces of the leaves better, adheres better and uses less actual sulfur (approximately 10 percent by volume). This reduces the impact on the soil.

Powdery Mildew

Podosphaera leucotricha

Powdery mildew overwinters in apple buds as fungal mycelia, tiny thread-like structures that produce white spores as the flower buds begin to open and shortly thereafter. It can affect the buds, flowers, leaves and fruits of the apple tree. Once infected, the flowers and new leaves will turn a powdery white color as the spores mature. These spores will be released to infect new areas.

The disease compromises growth and often kills the tissues. Affected fruits will often show russet netting over their surface. Powdery mildew grows best when temperatures are in the 66°F to 71°F (19°C to 22°C) range. Although it thrives in high humidity, it does not do well when it is wet, as the water washes away its spores. It is often called the dry disease. In fact, if you are dealing with only a few small trees, rinsing them well with water can often prevent infection. Be sure to do this in the morning so that

Powdery mildew differs from many fungal diseases because it prospers in dry conditions.

the foliage will dry and not cause other types of infections that like wetness. In general, drier areas are more susceptible to powdery mildew.

Choosing mildew-resistant apples is the best way to avoid this disease. Many of the newer cultivars developed for scab resistance are also mildew resistant. Having a good site open to sun and air circulation will help prevent the disease from establishing. It does better in low light levels, so try to keep your trees relatively open to sun. Avoid high-nitrogen fertilizers as rapidly growing foliage provides a good habitat for the disease. Use instead organic supplements such as compost and high-protein meals.

Commercial growers usually rely on fungicides that are used against apple scab to prevent powdery mildew as well. If you are looking for more environmentally friendly methods, here are a few proven suggestions.

Milk mixed at a ratio of one part milk to two parts water has proven very effective at preventing the disease. Apply just before silver tip and after full bloom. Reapply until early summer, when temperatures climb to levels inhospitable to mildew, above 71°F (22°C).

You can use common baking soda for prevention, mixed at the ratio of one tablespoon of soda, one teaspoon of dormant oil and one teaspoon of insecticidal soap or a pure soap (not detergent) per gallon of water. Slightly higher rates of soda can be more effective, but take care not to add too much as it can burn foliage.

Even ethanol-based mouthwashes have been effective at prevention. Use at a ratio of one part mouthwash to three parts water. Vinegar has also been used at the ratio of two to three tablespoons of a 5-percent acetic acid concentration per gallon of water. Again be careful with your ratios. Too strong a mix can injure foliage.

Once symptoms are noticed you can rely on commercial products that contain potassium bicarbonate. This is effective at killing the spores and, because it is absorbed into the surface of the leaf, can kill existing infections. There are also products that contain a bacterium called *Bacillus subtilis* for control once symptoms are noticed.

Sulfur, lime sulfur and Bordeaux mix (a blend of copper sulfate, hydrated lime and water) can also be effective, but be cautious as these can affect the foliage as well as the soil life. Sulfur should never be used in hot weather or it will burn foliage.

Flyspeck

Zygophiala jamaicensis

This fungal disease appears in mid to late summer on the surface of the apple. It affects most cultivars but is most noticeable on green- and yellow-skinned apples. The disease forms spots that resemble flyspecks grouped in clusters that can range from fractions of an inch to inches in diameter.

The damage done to the fruit is only aesthetic. It does no harm to eat affected fruit and does not cause breakdown in storage.

Commercial growers usually monitor apples in the interior of trees, where moisture levels remain highest. If spotted, a fungicidal spray is used to prevent its spread. As this fungus often grows on brambles, it may prove beneficial to remove brambles from the orchard. Regular pruning to keep the tree open to air circulation and more rapid drying will help, though climates with high humidity may still allow the fungus to grow.

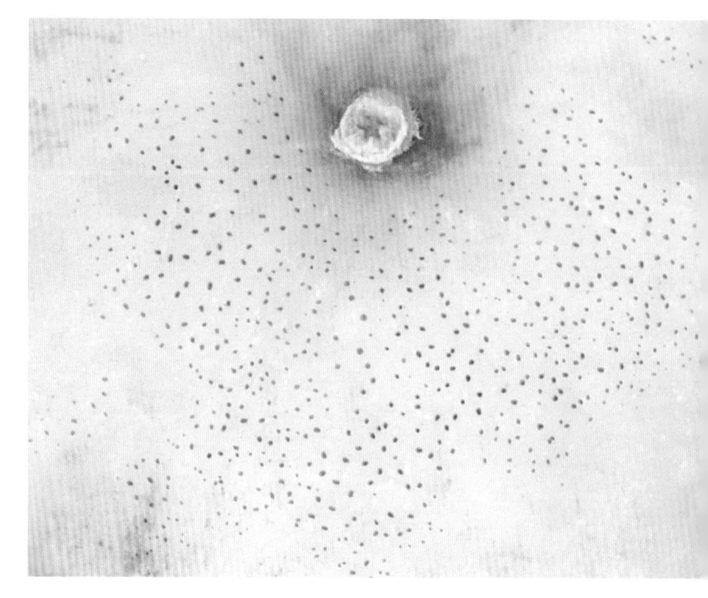

Flyspeck occurs late in the season, especially toward the interior of the tree.

If your fruit is affected, a simple wipe down with a damp nylon scrubber will clean the apple. Though a bane for commercial growers, this is not a serious disease for the home grower.

Sooty Blotch

Peltaster fruticola, Geastrumia polystigmatus and *Leptodontium elatus*

This disease behaves in much the same way as flyspeck and is often grouped together with it.

Rather than forming groups of spots, sooty blotch appears as smudges or olive-green spots

Sooty blotch is another late-season fungal disease that lives on the surface of the apple.

Canker is usually found where winter injury has occurred.

with irregular margins. It does appear as if the affected area of the fruit is covered in soot.

The biology of this group of fungi is similar to flyspeck, and its appearance on the fruit is also in mid to late summer.

The use of fungicide, whether organic or not, is still the tool of most commercial growers. A damp scrubber will likewise remove the disease from the apple. Similar to flyspeck, sooty blotch affects only the appearance and not the usability of the fruit.

European Canker

Nectria galligena

Cankers are usually introduced into the orchard from infected nursery trees or where winter injury or mechanical injury has occurred in a tree. Crotches in the trunk or at branching junctions are sites where water can sit for long periods and rot tissue. Injuries from snow load, ice or pruning activities can open wood and expose tissue. These are favorable spots for canker to develop. Though the cankers may not initially be apparent, the first signs will be reddish-brown spots near injury sites, leaf scars, spurs or pruning cuts. Gradually a sunken dark area will become noticeable. The canker will spread outward and on smaller stems may completely encircle the branch, killing the growth above it. Infections on larger branches or trunks usually cease after a year when the tree forms a callus layer along the edge of the infection.

This disease is not treated with fungicides, though fungicide use in the orchard may help control it. Pruning out infected wood is by far the most effective method of control. Where infection has occurred on a main trunk, clean out any destroyed tissue. Make a fine cut with a knife along the edge of the site to expose a layer of healthy cambium for future callus formation and paint the entire site with a latex paint.

Using winter-hardy cultivars will help minimize the incidence of canker. Tender trees have a higher incidence of damaged tissue, providing more entry points for the fungus.

Collar, Crown and Root Rot

Phytophthora cactorum

The disease is caused by a water-borne mold that affects the bark and outer layer of the roots. When water is allowed to sit at the base of the tree, infection can occur and cankers develop that will prevent the flow of nutrients to the roots. When this disease affects the scion-rootstock union, it is called collar rot; when it affects the area between the soil and the union, it is called crown rot; and when it affects the roots below the soil, it is called root rot.

Detection is difficult as the damage usually occurs beneath the soil surface, and the tree may not initially show any symptoms. Be on the lookout for yellow leaves appearing throughout the crown, poor vigor or purplish-colored leaves toward autumn. Once infected, trees rarely recover, and removing them is necessary.

This apple tree is suffering from crown and root rot.

This disease is usually confined to soils that are heavy and drain poorly, or in orchards where irrigation is allowed to saturate the soil around the crown. Choosing well-drained soils and providing drainage in heavier soils is critical for preventing crown rot.

Choosing crown rot–resistant rootstocks and cultivars is recommended, particularly if you have heavy soils. It has been reported that seedling rootstocks, as a rule, seem more resistant to crown rot, but as always every rootstock must be judged on its own merits.

Cedar Apple Rust (Juniper Apple Rust)

Gymnosporangium juniperi-virginianae

This disease requires the presence of eastern red cedar (*Juniperus virginiana*). When infected, eastern red cedar, actually a tree species of juniper, forms irregular galls on the stems. Depending on the maturity of the galls they can be brown, reddish brown or pinkish brown. During wet periods in spring, orange extrusions of jelly-like tissue, called telial horns, eject spores. Each time the horns are wetted a new extension is formed and more spores are ejected.

When these spores land on young apple leaves and fruit, and if there is a film of water

Cedar apple rust only occurs where eastern red cedar (*Juniperus virginiana*) grows.

present for an adequate amount of time, infection will occur, and orange pustules will form and grow on the upper sides of young leaves or on the fruit. These sites will eventually release spores into the air that will germinate on nearby eastern red cedar and complete the cycle.

This disease can seriously affect tree vigor and fruit quality.

Because the disease requires eastern red cedar for half of its cycle, removing such trees near the orchard is imperative. If this is impossible, fungicides will be necessary to prevent infection. In more northerly sites this disease is rarely a problem as eastern red cedar does not generally grow in Zones 4 and colder.

There are a number of apple cultivars that are resistant or immune to cedar apple rust.

Quince apple rust (*Gymnosporangium clavipes*), a similar disease, can also affect apple trees, but this disease requires the presence of tree quince, which is not hardy in most northern areas.

Bacterial Infections

Fireblight

Erwinia amylovora

This can be a devastating disease when not caught early. It is caused by a bacterium that enters the tree through the flowers or growing tips of shoots. Infected flowers will first appear waterlogged and gray but will quickly turn black. The stems will blacken and form a characteristic shepherd's crook or candy cane–shaped end and will form sites where bacterial ooze will drip and spread infection to new areas. If not pruned out early, fireblight will continue to move through the tree and can eventually cause its death.

Some rootstocks are highly susceptible to fireblight. Infection usually occurs when suckers emanating from the rootstocks become infected. If not caught immediately it will kill the rootstock and, thus, the tree. Be sure to remove rootstock suckers as soon as they are noticed.

The characteristic "shepherd's crook" shape and burnt look of a fireblight infection indicates immediate action is required.

Fireblight occurs most frequently on rapidly growing trees. In areas affected by fireblight, you should be cautious when using nitrogen-rich fertilizers. High levels of nitrogen will produce watery succulent shoots that are the most prone to fireblight infection. Using organic sources of nitrogen rarely produces such growth.

Fireblight is not usually a disease that occurs with any regularity in areas where temperatures drop below –31°F (–35°C). This is one instance where cold winters are an advantage. This should not make growers in such areas complacent. Mild winters can create conditions favorable to fireblight, and several mild winters in a row should make growers doubly cautious.

Detecting fireblight early is critical to control this bacterial disease. Cut infected stems out well below where symptoms show. It is highly recommended to sterilize the pruning equipment with alcohol between every cut. The cuttings should be burnt or buried.

Physiological Disorders

Bitter Pit

Bitter pit is a physiological disorder that results from low calcium levels in the fruit. While an adequate supply of calcium in the soil is recommended, this does not always eliminate bitter pit. It seems that weather conditions and other stresses can cause the tree to uptake less calcium or to transfer calcium in the fruit to other parts of the tree.

In dry years when water is in short supply, calcium, which moves slowly through the tree in the best of times, will take even longer to move through the tree. Fruit is a poor competitor for a tree's calcium, so much of it will move into the shoots, particularly when there are high nitrogen levels in the soil pushing vegetative growth.

Bitter pit is due to calcium deficiency during development.

Excessive levels of potassium and magnesium in the soils can also cause a slower uptake of calcium, even when there is sufficient calcium in the soil. Soil tests should be done to establish the levels of all these minerals.

Symptoms may not show up until harvest or, most likely, in storage. Affected apples form pits that are brown or gray. Beneath these pits the flesh becomes brown, dry and spongy, making them unmarketable.

This disease affects some apples more than others. It also seems to occur more on trees that are growing vigorously with a small crop of fruits.

Calcium sprays, such as calcium chloride or Cor-Clear®, will reduce the incidence of bitter pit, but these should be used with caution. Such programs begin spraying in July every

10 days. Calcium chloride should not have a concentration above 16.5 oz. per 25 gallons (500 g per 100 L) or severe burning of leaves and russeting of fruit can occur. It is recommended sprays be applied when temperatures are below 77°F (25°C).

Prevention of bitter pit includes keeping the calcium levels in the soil at adequate levels, avoiding high-nitrogen fertilizers and summer pruning vigorously growing trees, especially those with light crops.

Like bitter pit, watercore is due to calcium deficiency. Affected apples will break down in storage.

Watercore

Watercore is not entirely understood, but it seems that certain conditions can cause the vascular system of the fruit to deliver water that can no longer be held by the cells into the spaces between cells in the core and areas around the core, a process known as dumping. This saturates the flesh, giving it a waterlogged, translucent appearance.

This condition appears most frequently where orchards mature their fruit in high temperatures and under intense sunlight, but it can be a problem in cooler areas, particularly with certain cultivars.

As with bitter pit, calcium appears to play a major role in watercore. Calcium is important in the maintenance of cell membranes and the movement of materials between cells.

Although watercore does not adversely affect flavor, it does affect the look and feel of the flesh. If stored, apples affected by watercore will eventually break down and begin fermenting, though slight watercore can dissipate in storage.

The best methods of prevention include avoiding high-nitrogen fertilizers, especially on lightly cropping trees, maintaining good levels of calcium in the soil and pruning vigorous trees in the summer. Some growers spray calcium to prevent both watercore and bitter pit.

Cork Spot

This condition shows up as small, depressed areas on the fruit's skin. The depressions resemble a mark that a soft tap with a ball-peen hammer might make. Under the skin, a corky layer forms that renders the apple less palatable. This condition is caused by a deficiency of calcium. It is similar to bitter pit.

Maintaining a neutral pH will help alleviate this condition, but even with a proper pH the fruit may not receive enough calcium. Often this occurs when there is vigorous growth in the tree, particularly when water sprouts are left to grow after a hard pruning. The calcium is diverted to growth areas, and the fruit suffers

Cork spot is yet another result of calcium deficiency. It looks as if the apple was hit lightly by a ball-peen hammer.

correspondingly low levels of calcium. Calcium sprays are available to alleviate this condition, and removing water sprouts will also help.

Other Pests

Deer

While various animals, such as porcupines and bears, can damage apple trees, these instances are the exception. Deer are the rule. This is one animal that can cause major damage to your trees. When a significant population of deer moves into your neighborhood, you will need to take action. Deer love apples, but they particularly like to browse on the soft new growth found at head height — deer-head height, that is. Newly planted trees provide such a treat, especially when you have fed them, watered them and adjusted the soil. Deer can sniff out where growth is most robust. Your young trees are one, two, perhaps three years old and maybe just coming into production. So tasty. Coming upon the aftermath of such a dining experience is disheartening.

One final and simple answer to the deer issue is a fence. It also can be a rather expensive answer, depending on the size of land that needs protection. Fences have to be at least 6 ft. (1.8 m) high, but 8 ft. (2.4 m) is recommended.

All in all, a tall fence is an investment that has to be justified by your needs and desires. The upside is that if it's a well-made fence with rot-resistant posts you won't have to replace it for a long time. And you won't be running after deer, applying repellents every month or worrying at night. If deer get inside your fenced area it's either because someone left the gate open or you didn't build it high enough.

If investing in a fence is out of the question, there are many deer repellents on the market. These are usually blood-based, as deer tend to shy away from the smell. The repellents require reapplication after rain, when they have been washed off. A few have more residual staying power, nonetheless you will have to reapply the repellents often. One possible downside with relying on repellents is that deer are clever and will eventually realize your efforts are not really an impediment to

Deer love the new shoots on apple trees. Once bitten, regrowth is slowed by chemicals in the deer's saliva.

their dining pleasure. We have had the same experience with spreading hair of all types — from human to lion (we have a zoo nearby). It works only for a while. Some growers swear by certain bars of soap.

The best repellent I have ever tried is male human urine. Mix one part urine to 10 or so parts water, and spray the perimeters of your orchard. It takes little time if you use a back-pack sprayer with a hand pump. If there is heavy feeding pressure in certain areas of the orchard you can spray the tips of the trees. This is where they will be feeding, and it will do no harm to the growing tips. Despite some people recoiling at the thought, it is easy to source, barely smells when sprayed and is effective at marking your territory. However, use this repellent quickly or refrigerate it. You don't want it sitting around in a warm spot — then it will smell. I think it makes the best repellent, but it comes with all the downsides of the others, like keeping up with applications, particularly after rain.

The right kind of dog is also an excellent deer repellent. Deer are wary of dogs that show territorial aggression. You want to make sure your dog does not chase the deer beyond your property as the aim is not to have your dog kill the deer. Just having a dog that is around the orchard seems to be helpful.

Damage done by the yellow-bellied sapsucker can kill portions of the tree above its work.

Small Mammals and Birds

There are a host of other pests that can do harm to apple trees. Perhaps the most dangerous to young trees are field mice and voles, who feed on the tender bark. Most damage by these rodents occurs in winter, often unbeknownst to the grower, as feeding can be hidden under the layer of snow. The easiest way to combat these critters is to use tree guards around the trunks. Plastic tree guards are the easiest to use, but a wire mesh guard made of 0.25-in. (6-mm) hardware cloth is also effective and will last for many years. Be sure the guard is dug into the soil a few inches, as the voles can do damage just below the soil surface. Tree guards should be checked frequently as pests such as the appletree borer can infest the trunks without the grower noticing. Consider removing the guards from April to October.

Another animal that can cause severe problems is the yellow-bellied sapsucker. This member of the woodpecker group drills holes in nearly perfect geometric grids through the bark. They return later to drink the sap and eat the insects that have been attracted to the sap. This can cause severe disruption to the movement of sap in the tree and eventually the death of the section above the damage.

7 Harvesting and Storing Apples

The Harvest

After the work of planting, fertilizing, weeding, pruning and preventing pest damage and diseases comes the activity the grower has been anticipating all season: the harvest. Harvesting apples is a task that involves knowing when the optimal pick date is reached, organizing the actual picking, ensuring the apples are picked with care and understanding the needs of each cultivar before entering storage.

Picking Date

Biting into an apple is one way to test its ripeness, but even though an apple might taste good, it may not be the ideal time to pick. Each cultivar has a period of time, usually just a week or so, when the fruit is at its optimal stage of development for peak flavor and/or storage. Picked too early, an apple may taste "green" or "grassy" or have high acidity. Immature apples will also not store well. Picked too late, the apples may be mealy and/or not keep well in storage. If picking is delayed you may also lose apples from premature drop. It should be noted that if you are storing the apples in a controlled atmosphere, they may have to be picked slightly earlier than peak flavor.

◂ A harvest of apples — the reward of hard work and nature's bounty.

After Apple-picking

Robert Frost

My long two-pointed ladder's sticking through a tree
Toward heaven still
And there's a barrel that I didn't fill
Beside it, and there may be two or three
Apples I didn't pick upon some bough.
But I am done with apple-picking now.
Essence of winter sleep is on the night,
The scent of apples: I am drowsing off.
I cannot rub the strangeness from my sight
I got from looking through a pane of glass
I skimmed this morning from the drinking trough
And held against the world of hoary grass.
It melted, and I let it fall and break.
But I was well
Upon my way to sleep before it fell,
And I could tell
What form my dreaming was about to take.
Magnified apples appear and disappear,
Stem end and blossom end,
And every fleck of russet showing clear.
My instep arch not only keeps the ache,
It keeps the pressure of a ladder-round.
I feel the ladder sway as the boughs bend.
And I keep hearing from the cellar bin
The rumbling sound
Of load on load of apples coming in.
For I have had too much
Of apple-picking: I am overtired
Of the great harvest I myself desired.
There were ten thousand thousand fruit to touch,
Cherish in hand, lift down, and not let fall.
For all
That struck the earth,
No matter if not bruised or spiked with stubble,
Went surely to the cider-apple heap
As of no worth.
One can see what will trouble
This sleep of mine, whatever sleep it is.
Were he not gone,
The woodchuck could say whether it's like his
Long sleep, as I describe its coming on,
Or just some human sleep.

Knowing the proper pick time for the apples you grow is a must. Talking to local growers can be very instructive, as their livelihood depends on this knowledge. In terms of what the fruit can tell you, the color of the seeds is a good indication of ripeness. Unripe apples will have ivory-colored seeds, but a ripe apple will have brown seeds. There are also starch iodine kits that can provide a reliable guide to picking times. By placing the solution (composed of 4 percent potassium iodide and 1 percent iodine) on the cut surface of an apple, you can measure the amount of starch in the apple. As starches diminish, which happens at maturity, the intensity of the purple/black color lightens. There are several guides online to instruct you about this, but be aware that different cultivars can show slightly different results.

Unfortunately you cannot rely on a book like this to tell you the picking date for your cultivar of apple. The pick date in Geneva, New York, will not be the same in a more northerly orchard. Peak flavor may not be reached for some cultivars until the apples undergo a period of storage, but the proper pick time is still vital. For example, Golden Russet will not taste the way you expect at its proper pick date. These apples develop their desired taste in storage after one to two months.

A picking bag is indispensable for harvesting any amount of fruit.

Organizing the Harvest and Picking

Even if you are only picking a tree or two, a little organization helps. First, you must have the proper equipment. If your trees are above head height, you will need something to stand on. This might be a stepladder or, if the tree is taller, an aluminum orchard ladder with an adjustable third post. Picking a tall tree can be dangerous, so be sure your support is reliable.

You will need something to hold the harvested apples as you pick. Ideally you will have a fruit picking bag that is strapped to your chest. This frees your hands as you move up and down the ladder and prevents damage to the fruit because you can gently release the

Picking bags allow for the gentle movement of fruit into a bin.

contents of the bag into a larger container. Fruit pickers attached to a pole can be quite handy to get at fruits that are out of reach.

The actual method of picking of the apples is vital to maintaining quality. It is said that a basket of apples with lots of leaves is a sign that the apples have not reached their proper picking time. Be sure that apples are only picked at the proper maturity.

Training a picker on how to properly pick an apple will prevent damage to the tree and the crop. Most apples can be released easily if they are rolled upward so that they snap off the spur. Pulling apples can result in harm to the spurs and often causes nearby apples to fall.

Apples can bruise during picking, but more often this occurs when they are put into larger bins or when the bins are moved. It is a real shame to go through all the efforts required for a good crop only to have a large portion damaged by improper handling, so be sure to take care while picking and transporting your apples to storage.

Aftercare and Storage

Once picked, apples need to be handled gently and stored where proper temperature and humidity will keep them in the best condition. Most apples can be moved quickly into a storage area that will maintain a temperature of approximately 37°F to 40°F (3°C to 4°C). Keeping them at temperatures above 50°F (10°C) for a prolonged period will hasten internal breakdown and make them unsuitable for anything but cider. High humidity is also important. The average refrigerator has a relatively low humidity level, so keeping the apples in plastic bags with a few small holes is the best option. In commercial storage units humidity levels of 80 to 95 percent are maintained when storage temperatures are 30°F to 40°F (–1°C to 4°C). The lower the storage temperature, the higher the humidity level should be. Some apples, such as Golden Russet, will shrivel quickly if kept at low humidity levels. If you are keeping apples in refrigerated storage, be sure there are no apple trees or scion wood kept with them, as the maturing fruit releases

A room in a controlled-atmosphere storage facility.

ethylene gas, which, in high concentrations, can kill the buds within days.

There are some apples that need a postharvest period at temperatures of 50°F (10°C). Honeycrisp is a good example. If placed into refrigerated storage immediately, it can develop physiological conditions such as soft scald and soggy breakdown, which affect the skin and eventually the flesh. There is quite a bit of information online about the best postharvest conditions for at least the most common apples.

Commercial storage units often have sealed rooms to keep apples for long-term storage, and these are only opened when the apples are sold. These rooms have extra carbon dioxide pumped into them so that they maintain CO_2 levels of 2.5 to 4.5 percent and oxygen levels at 2.5 to 3 percent. For most apples these conditions are ideal for long-term storage; however, there are exceptions. Again, Honeycrisp has had problems in controlled atmosphere rooms, though some research has shown that you can successfully store Honeycrisp with proper postharvest temperatures followed by storage in a room with a carbon dioxide level of 1.5 percent and an oxygen level of 3 percent. Obviously such details are only applicable to large-scale growers. Most of us do not have access to such storage facilities, so keeping the temperature and humidity as close to the proper level as possible is the best we can do.

8 Grafting and Budding

Passing on the Legacy

The origin of every fruit tree begins with a seed, swelling with moisture, at its core a tiny grouping of cells, the embryo. Just as a crystal of any element grows in a determined pattern, so does the embryo. The apple has a large genome, meaning that even when the seeds of a desirable tree are grown out their fruit will rarely resemble the mother, but an apple of no significance might produce a tasty sensation as well. In other words, there are endless possibilities.

Though apple flowers contain both male and female parts, the flowers are nearly always self-sterile, so any seed produced will be a combination of the mother tree and another, which donates its pollen. The possible combinations from these unions are nearly infinite. Though a seedling might resemble one of its parents, chances are it will be quite different. It is the latest expression of the apples that came before it, a unique combination that will determine the tree form, branch angle, bark pattern, flower color and the look and taste of its fruit.

◂ A bud on a cleft-grafted scion emerges through Parafilm® wrap.

Like all living things, fruit trees grow for a time, then die. It is food for thought to imagine all the fantastic fruits that have

Grafting is the transfer of a desirable cultivar onto a rootstock — a perfect marriage of science, craft and wonder.

grown on trees that no person ever saw, let alone tasted. Throughout history some of those special seedlings have been noticed and their fruit harvested, but of course such trees were individuals that eventually died, their unique offerings vanishing with them. It is fair to say that most of these trees were found by people of the forests or fields, people intimately connected to the sites where these trees took root.

Some time ago, perhaps three to four millennia, most likely in China, someone discovered a trick, an art, a gift. This person took likely a knife and mechanically joined a piece of one tree onto another and performed the first graft. Since that moment, we have been able to keep trees alive far beyond their normal life spans, sometimes for centuries. This was the beginning of grafting — the revolution that marked the start of modern fruit culture.

Grafting involves taking wood from a tree you wish to propagate and mechanically joining it to another with roots, most often a young seedling. Once united, water and nutrients flow from the rootstock into the graft wood, called the scion, and growth can begin. It is probably safe to say that the first grafts were considered magical, and, I assert, grafting remains so. The idea that you can slice a piece of wood from one tree and attach it to another and have that small piece turn into a new tree with identical characteristics to the original seems outlandish and astonishing. Even today, in a time when "miraculous" developments seem commonplace, grafting evokes awe in those who first witness it, and even in those who practice it as part of their everyday livelihood.

Grafting is both an art and a science. The art involves becoming adept at shaping both scion and rootstock so that the union is so tight that the cells of both press tightly against each other. Watching an accomplished grafter is like watching any talented craftsperson. Despite being a simple process, it takes several hundred or even several thousand attempts to become fluid and quick with the cuts. Though many try, some will lack the patience and dexterity to become good at the task.

The science of grafting involves understanding how growth occurs within a tree or shrub. The outside of a branch is covered by a corky layer we know as bark, which protects the inner portions from injury by wind, fungi, bacteria and insects. Beneath the bark lies the cambium layer, where growth in a tree takes place. This thin sheath of cells carries the sugars produced by the leaves. It transports these sugars throughout the upper portions of the tree and down to the roots and provides the energy that enables the cambium layer itself to expand.

A graft's success depends on positioning the cambium layers of both scion and rootstock close enough that the cells can divide and multiply to fill any spaces between the two pieces before the scion runs out of food and water, placing them in such close contact that water, sugars and nutrients can flow between the pieces. A graft that misaligns the cambiums or leaves too much airspace between the two is doomed to fail. The marriage of art and science is what allows us to grow thousands of identical trees. It is the reason you can buy a McIntosh apple or a Clapp's Favorite pear today, decades or even centuries after the original seedling has perished of old age. Though the process can be described and understood, it remains wondrous.

Own-rooted Trees

There are methods of producing apple trees on their own roots, rather than by conventional budding or grafting.

The advent of tissue culture has made possible the production of trees in a laboratory setting. Small amounts of tissue are harvested from sterilized material and grown in a gel-like medium called agar that contains various nutrients and growth hormones. The tissues begin replicating, a process called proliferation. These replications are divided in a continual process to produce as many as necessary. To produce roots on this vegetative material, the small replications are placed in a different medium that has growth hormones that trigger root production. Once rooted, these "plantlets" are brought out of the test tubes or growth jars to begin the process of acclimation to fluctuations in temperature, humidity and air movement. Care must be taken to rogue any off-types that may occur.

There is a second method that involves taking lengths of the current year's growth, usually at leaf fall, and placing them standing upright in bins filled with moist sawdust. Both air temperature and bottom temperature are regulated, the basal temperature being higher than the air temperature above the cuttings. The bases of the cuttings are treated with indole-3-butyric acid (IBA), a hormone that induces rooting. Once rooted, these cuttings are planted outside. This method requires meticulous attention to detail.

Hugh Ermen (1928–2009), an important breeder in the United Kingdom, researched own-rooted trees and found great merit in them. Below are some advantages and disadvantages he noted:

Advantages

1. There are no problems of incompatibility between the scion variety (cultivar) and rootstock. It has been shown that rootstocks may have different needs in terms of nutrients and that they may not deliver the levels of nutrients required by the scion, thereby reducing tree health.
2. Varieties (cultivars) can come into growth and flower earlier than the rootstock, which can lead to poor fruit set. Own-rooted trees often exhibit better fruit set.
3. Trees on their own roots can have better fruit quality and storage life.
4. Own-rooted trees may exhibit better disease and insect resistance.

Disadvantages

1. Own-rooted trees will exhibit their natural vigor, which will not be controlled by the rootstock.
2. There is very little research on, and production of, own-rooted trees.
3. Many are biased against own-rooted trees.

There can be little doubt that trees grown on their own roots have many advantages. In the future we may see more orchards set out with own-rooted trees. The problems of tree vigor will need to be addressed cultivar by cultivar and soil nutrition and pruning methods adjusted accordingly. For the moment, however, grafting still remains the most common method of producing fruit trees.

The one disadvantage this author sees with own rooting is that many cultivars will not be as productive or as hardy on their own roots. Specific rootstocks can provide dwarfing or vigor, precocious bearing, heavier bearing and hardiness. Especially in cold-climate orchards, grafting and budding will likely remain the most important propagation method for the foreseeable future.

Rootstocks

By far the largest percentage of apples grown are grafted or budded onto rootstocks. The grower must ensure the chosen rootstock will be hardy for their area and assess which one best suits their site, soil, cultivars, level of maintenance, capital costs, climate and availability of labor. Beyond this, most rootstocks are chosen for their ability to confer productivity and precocity to a cultivar.

Dwarfing rootstocks induce early and prolific fruit production. Because the energy of

A bundle of rootstocks ready for grafting.

the plant is directed to seed development rather than vegetative expansion, the tree remains smaller. Most dwarfing rootstocks will create trees 25 to 40 percent the size of a standard tree. The advantages of small trees are numerous. Picking can be done from the ground or with a stepladder, which is far safer than having to balance ladders on the limbs of tall trees. Managing trees is easier, as all parts can be inspected from the ground, and therefore problems can be dealt with more quickly. Pruning and thinning of fruit is easier and safer.

There are some disadvantages to small trees. Dwarf trees often need staking to keep them upright, as many of the rootstocks are poorly anchored. A tree that needs continual support will require a system that can be costly to install and maintain. It is also true that the volume of productive area in a dwarf orchard is not as great as that in a more standard orchard, though there are factors, such as consistently larger fruit size, that can offset this.

There are many rootstocks that are semi-dwarfing. These generally produce trees that are between 40 and 60 percent of the standard size. Semi-dwarfing trees are often better anchored than dwarf trees, and many do not require long-term staking. It should be remembered that staking a new tree for up to a year is advisable no matter the rootstock.

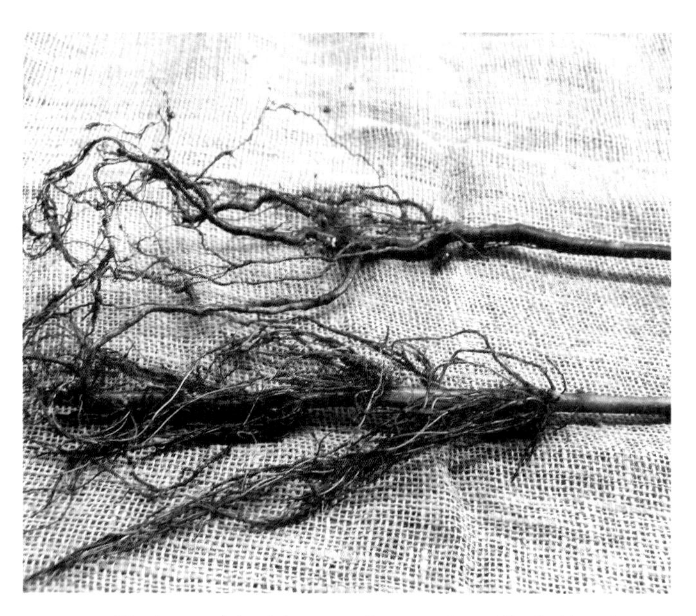

The top rootstock is grown from seed, and the bottom is a rooted shoot from a clonal stool bed.

In northern areas, hardiness becomes a critical issue. Many well-known dwarfing rootstocks are unreliably hardy in cold climates. Some, under normal climatic conditions, may perform well in colder areas, yet when unusual conditions occur, they may suffer severe or catastrophic winter injury. As an example, the winter of 1980–81 was just such a test in Atlantic Canada, northeastern New England and Quebec. Temperatures in February climbed to unseasonably high levels of 50°F to 60°F (10°C to 15°C) and remained so for nearly a week before plummeting to –4°F (–20°C) very quickly. Trees grafted to more tender rootstocks had come out of deep dormancy during the long warm period and were unable to reenter dormancy quickly enough when the cold returned. Many new plantings, most of which were grafted onto dwarf rootstocks from England, died. Older orchards that had cultivars that were grafted onto hardy seedling rootstocks fared far better, as these roots required a longer warm period to break dormancy, a protective adaptation of trees coming from cold climates.

That being said, there is a whole new generation of rootstocks coming from places such as Geneva in New York State, Russia, Vineland in Ontario and Poland. Many would make excellent rootstocks for the north.

Rootstocks are of two general types: seedling and clonal. We shall explore both types.

Seedling Rootstocks

Seedlings generally offer better longevity, particularly in colder regions. They impart more vigor and form a strong, deep root system. A cold hardy seedling will generally shut the cultivar down earlier in the season, and this helps prevent early winter injury. These seedlings usually come out of dormancy later and can slow bud development of the cultivar in spring, thereby preventing premature growth that can be injured by late freezes. While there are many new hardy clonal rootstocks becoming available, seedling rootstocks have, at least in the past, been the roots of choice for the coldest areas.

If rootstocks are seed grown, there will be variability among the seedlings. Characteristics such as hardiness, productivity, precocity and vigor may differ. This diversity, however, can be an advantage in the orchard. Seedling rootstocks may have different levels of susceptibility to factors such as weather or disease. Certain cultivars of apples have been recognized as producing hardy seedlings with good uniformity, and if possible these seeds should be sought out.

Seed for rootstocks is most often collected from specific species or cultivars with known characteristics. The apples from these mother trees should be harvested at maturity. Most apple seeds will be whitish when unripe, then will turn tan or brown when ripe. Once the apple is picked, separate the seed from the flesh by crushing the fruit either with hand tools or by various mechanical methods. The goal is to create a fine mash yet do no damage to the seed. When the resulting mash is put in water the good seeds will sink to the bottom, while the empty seeds, flesh and core materials will float to the top, where they can be skimmed off. After several skimmings, place the remaining seeds on fine screens, absorbent cloths or papers to dry. When the seeds are dry enough to handle sow them directly into prepared seed beds that have been enriched with compost and adjusted to a neutral pH. The seeds should be planted approximately 0.4 in. (1 cm) deep and lightly covered by a fine mulch or sand approximately 0.4 to 0.8 in. (1 to 2 cm) deep. Apple seeds need to be moist and cold for at least 90 days before they germinate, so planting them in the late autumn/early winter period satisfies this requirement.

Germination will occur in early to mid-spring. Because apple seed germination is usually not 100 percent successful, you might want to place a seed every 2 in. (5 cm) and later thin the seedlings to 4 in. (10 cm) apart in spring if necessary. With good growing conditions (adequate soil fertility, available water and summer temperatures), the rootstocks should be ready to harvest at the end of the second growing season. The ideal stem size for grafting is 0.2 to 0.25 in. (5 to 6 mm) in diameter.

Apple seeds collected from Beautiful Arcade, a cultivar known to be a source of desirable rootstocks.

If growing the seedlings in containers, the container should be approximately 2 in. by 2 in. (5 cm by 5 cm) and 3 to 4 in. (7.5 to 10 cm) deep. Sow the seeds in a seedling mix that has excellent drainage but will retain moisture. The mix should be pH adjusted to between 6 and 6.5, and there should be a sufficient volume of material to allow adequate root growth. Place the containers in a holding area where moist soil and cool temperatures (34°F to 38°F or 1°C to 3°C) will provide the proper dormancy-breaking conditions, a process called stratification. The stratification period required for good germination is usually 90 days. Once this requirement has been met, move the containers to a warmer area and germination should occur within a few weeks. The roots of new seedlings should be kept moist but not wet, and the seedlings should have access to full sun. When the roots reach the edges of their pots the seedlings should be transplanted into larger containers or into the field for finishing.

Apple seeds that are not used immediately should be stored in sealed polyethylene bags and kept frozen, ideally at 25°F (–4°C). Seeds held in these conditions will be viable for several years.

Seedlings from any apple tree can be used for rootstocks; however, compatibility, performance and hardiness will vary widely. Seedling rootstocks are most often grown from seeds that are obtained from fruit harvested off cultivars known to produce seedlings with uniform characteristics. Sometimes seeds are collected from trees of a specific species, for example, *Malus baccata* (Siberian Crab), but the hardiness of such seedlings may vary depending on the geographic origin of the individual trees used as the seed source. Below are several seed sources that have proven suitable for producing reliably hardy rootstocks. I have given the average hardiness of the rootstocks, but be aware there is always the possibility that, with the tremendous genetic diversity in apples, the occasional seedling may be less hardy.

Hardy Seed Sources

Anis: This is a Russian cultivar whose seedlings produce moderate-sized trees. It has been used for many years in Russia and to a limited extent in the colder areas of the United States and Canada. Hardy to Zone 2.

Antonovka: This is a Russian cultivar whose seedlings have been used in northern areas for many years. This cultivar tends to produce seedlings with strong tap roots. Trees on this rootstock will average 70 percent of standard size. Although hardy in most sites, it has not proven as hardy as *Malus baccata* in the most extreme situations. Hardy to Zone 2.

Beautiful Arcade: Seedlings of this hardy Russian cultivar tend to produce large, early, sweet fruit. The seedlings are generally uniform, precocious and productive. Most cultivars grafted to Beautiful Arcade seedlings average 65 percent of standard size. The root systems are very fibrous and well anchored. Hardy to Zone 2.

Borowinka: Seedlings of this older Russian cultivar have been used successfully in many northern areas. It produces productive trees, though it is not considered hardy enough for extremely cold areas. Hardy to Zone 3.

Dolgo: Seedlings of the crabapple Dolgo produce vigorous trees with good productivity. The rootstock has good compatibility and superb hardiness. Hardy to Zone 2.

Malus baccata: This is considered the hardiest of all apple rootstocks. Its main disadvantage is graft incompatibility with some cultivars, so it is only recommended for extremely cold sites. Hardy to Zone 1b.

An orchard of Beautiful Arcade and Antonovka, two excellent sources of hardy rootstocks.

***Malus Columbiana* (Columbia)**: This crabapple species is often used in extreme northern areas. It has good compatibility with most cultivars, but it is very vigorous and this may delay the onset of fruiting. Hardy to Zone 2.

Malus Prunifolia: A very hardy Chinese species with plum-like foliage. It has a fibrous root system, and some growers report excellent productivity. Hardy to Zone 3.

Malus Ranetka: This is a cross between Dolgo and *Malus baccata*. The rootstocks produce vigorous and hardy trees, and they have shown good compatibility with cultivars. Hardy to Zone 2.

Selkirk: This is a pink-flowered crabapple whose seedlings have been used to a limited extent, particularly in the production of flowering crabapples. The rootstocks produce medium-sized trees. Hardy to Zone 2.

Clonal Rootstocks

Clonal rootstocks are produced in order to maintain the characteristics that make them valuable to the grower. The advantage of such roots is that their hardiness, productivity, precocity and vigor can be predicted from past experience. The disadvantage is that any weaknesses the clonal rootstock may have, such as disease susceptibility or lack of hardiness, will be transferred to any tree grafted to the rootstock.

Clonal stool beds are cut back each year to encourage strong new growth.

Most clonal rootstocks are produced in what are called stool beds. Existing rootstocks are planted out in rows in early spring. Once established, the plants are cut back to near the crown. As the vigorous new shoots begin growth, they are covered with a mixture of soil and sawdust that is kept moist and well fertilized. The portion of the shoots buried in the mix will form small roots. In the fall or early spring, the mix is brushed away and the lightly rooted shoots, called layers, are cut from the mother plant, and the process is repeated. The layers are usually planted out and grown for one year before being used for budding or grafting with the cultivars of choice.

Clonal rootstocks can also be produced from root pieces. Roots are collected in the late fall or very early spring, as soon as they can be dug. In this method, 4-in. (10-cm) long sections of root are cut from mature plants. Roots should be at least 0.25 in. (6 mm) in diameter. Once the pieces are cut they are stood up in a rooting medium, such as sand and peat. The end closest to the crown of the plant (the proximal end) is placed at the surface of the soil, and the end farthest from the crown (the distal end) is buried. Temperatures are maintained at 60°F (15°C). Once shoot formation occurs, the most vigorous shoot is chosen and any others are removed (if the plants are destined for stool beds all the shoots can be left).

If you have the proper facilities, removed shoots can be rooted. As these are coming from the roots, they are in a juvenile state, which makes them much more likely to root than a normal apple shoot, which is in an adult state. The very bases of the shoots should have a rooting hormone applied, and the shoots should be stuck in a rooting medium such as sand and peat or perlite and peat. Keep the humidity high until roots form. The shoots with roots can be transplanted either to containers or to the field for growing.

As noted above, apples generally do not root well from stem cuttings; however, some producers are using stem cuttings to multiply both cultivars and clonal rootstocks. The most common method is to take 20-in. (50-cm) sections of one-year-old wood in the fall. The bases of the cuttings are treated with rooting hormone and placed in a moist medium such as peat or sawdust. The portion of the cutting above the rooting medium should be

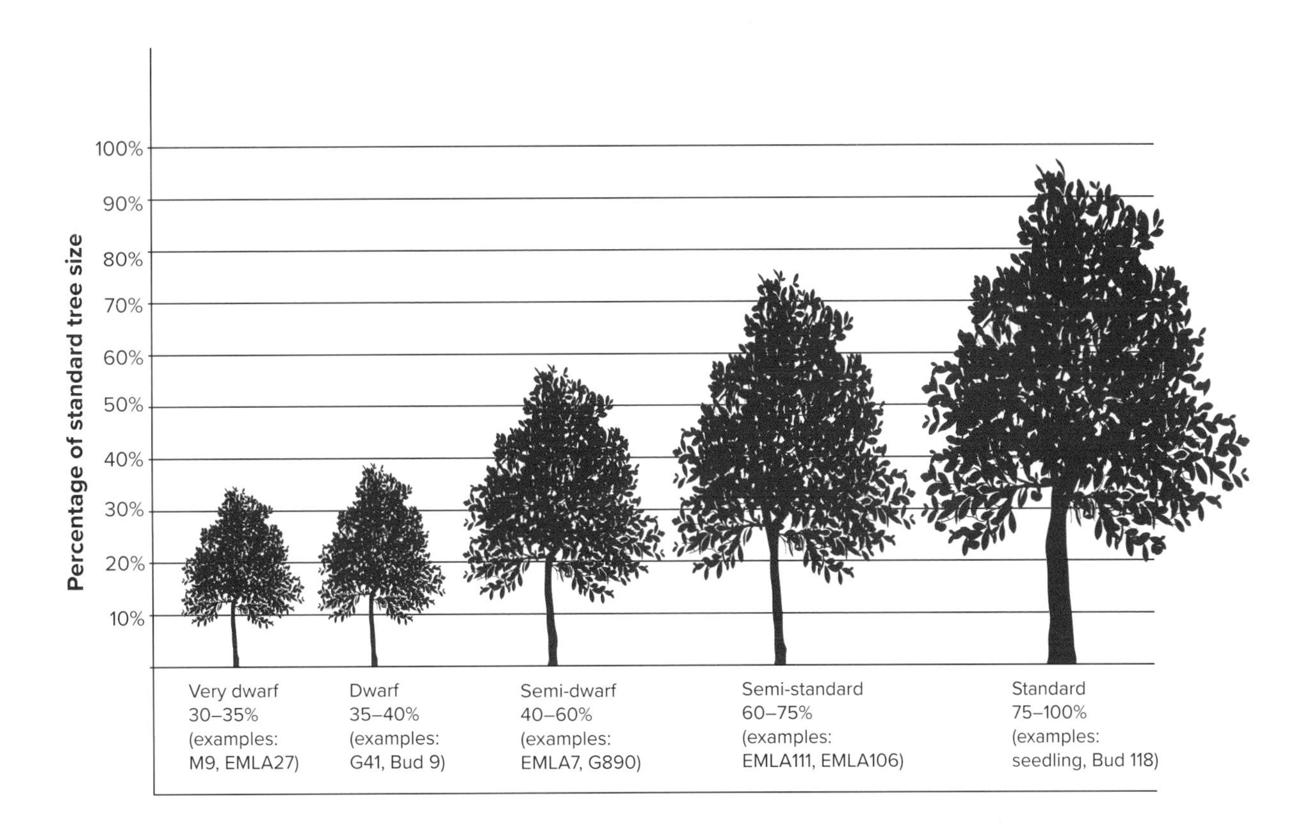

kept humid enough that it does not dehydrate, and temperatures should be kept just above freezing, while the basal area is kept at approximately 60°F (15°C) using a source of heat such as heating cables. Roots are encouraged to form well before the dormant buds begin to grow. This is a technique that requires strict attention to detail and vigorous, well-grown shoots. Not all cultivars respond well to this treatment, though research continues to make this method more economically viable.

A more recent production technique is tissue culture. Here the rootstocks are propagated in the same manner as described in the own-root section (see pages 105–106).

Clonal rootstocks are used extensively throughout the apple growing regions of the world. The most commonly found rootstocks in the Americas and Europe today are the Malling series. These were developed in England. The most famous of these rootstocks is M9, a very dwarfing rootstock with excellent production characteristics. This rootstock actually dates back to the 19th century in France, where it was called Jaune de Metz. It was brought to England along with several other older dwarfing roots. Beginning in 1917, scientists at the East Malling Research Station began the process of separating this confusing mishmash of often incorrectly named plants into distinct cultivars and giving them numbers with the prefix "M." Later, the John Innes Horticultural Institute in London, England, joined up with the Malling Station with the goal of creating rootstocks resistant to the woolly apple aphid. They crossed some of the Malling selections with the aphid-resistant cultivar Northern Spy to produce a series called the Malling-Merton rootstocks, designated by the prefix "MM." A later group was created that had many of the viral contaminants removed. These were given the prefix "EMLA." While the Malling series has proven to be an excellent group of rootstocks for the more temperate regions of the world, most are tender in very cold climates.

Because the Malling rootstocks have become so widespread, other rootstocks are often compared to them to give an idea of a rootstock's dwarfing capability. The smallest is M27, which produces a tree just over 20 percent of standard size. M26 produces a tree approximately 30 to 40 percent of standard, M9 about 40 percent and MM111, one of the hardiest of the group, about 65 percent.

Listed below are some of the newer introductions. Be aware that lists such as these are often out of date by the time they are printed. New rootstocks are being steadily introduced each year. Many of the selections below are more disease-resistant than earlier selections, and hardiness is being bred into many of the breeding lines. The future will probably offer many more hardy rootstocks.

Newer Hardy Clonal Rootstocks

Bud 9 (Budagovsky 9): This is a Russian rootstock developed by Dr. Budagovsky of the Michurinsk State Agrarian University. It is a cross between M8 and Red Standard, a Russian red-leafed apple. Its leaves are red, so when a graft union does not work, the shoots coming up from the root are literally like a red flag. It produces a dwarf tree similar in size to M9 (40 percent of standard). It is very hardy and resistant to collar rot and, to a lesser extent, apple scab and powdery mildew. It is, however, susceptible to fireblight. This rootstock should be supported with stakes. Hardy to Zone 3b.

Bud 10 (Budagovsky 10): A cross between Bud 9 and a seedling selection, this rootstock produces a small tree that is hardy, precocious and only 30 to 35 percent of standard size. This selection is resistant to fireblight and easy to grow in the nursery. It looks very promising. Hardy to Zone 3.

Bud 118 (Budagovsky 118): This selection is extremely hardy and is a good rootstock for the coldest areas. Its roots can withstand soil temperatures of 0.4°F (–18°C) and survive. The average apple root will die at 21°F (–6°C). It produces a freestanding tree that grows large. It is productive, precocious and resistant to fireblight. Hardy to Zone 2a.

EMLA26 (Malling 26 or M26): Some consider this the hardiest of the Malling series of rootstocks and is rated for –40°F (–40°C). Although it is not recommended for areas colder than Zone 4a, those in Zone 3b might want to trial it. It produces a tree 30 to 40 percent of standard. EMLA26 is extremely precocious, and it is recommended that any fruit be removed during the first three years to encourage better root development and a good central leader.

By season's end these vigorous new Bud 118 shoots have formed roots. The soil is removed, and the rooted shoots are cut for sale or propagation.

This rootstock should be staked. EMLA26 will runt out in very dry soils and does not perform well in wet soils. It will often form burr knots and should be planted with the graft union just above ground level to discourage their formation. It is susceptible to fireblight, so suckers should be removed immediately to prevent infection. Hardy to Zone 4a.

EMLA111 (East Malling-Long Ashton 111): This is one of the Malling-Merton series of rootstocks. It is a "cleaner" strain of the original MM111 that has had viral contaminants removed. While many of these rootstocks have not proven reliably hardy in colder areas, EMLA111 has been hardy to at least Zone 4, though it cannot be safely recommended for colder areas. It is semi-dwarfing, producing a tree approximately 65 percent of standard. Trees on EMLA111 are very productive and do not require permanent staking. Hardy to Zone 4a.

G16 (Geneva 16): The G Series was developed at Cornell University in Geneva, New York. G16 is a cross of Ottawa 3 and *Malus floribunda* (Japanese Flowering Crabapple), and it appears to have good hardiness for any but the coldest areas. It is similar in size to M9 (40 percent of standard size) and, like most small trees, will require staking and careful attention to weeding and feeding. G16 is immune to fireblight and collar rot but susceptible to woolly aphid. It is also sensitive to latent viruses in the chosen cultivars, so scion wood should be from virus-free material. It propagates well in stool beds. Hardy to Zone 3b.

G41 (Geneva 41): Another Geneva selection, G41 is a cross between M27 and Robusta 5. It is a promising selection that is slightly smaller than M9 with excellent productivity and efficiency. It induces good fruit size and is immune to fireblight and collar rot. G41 does not have the sensitivity to latent viruses that G16 does. This looks like a good replacement for M9. Hardy to Zone 4a (possibly 3b).

G890 (Geneva 890): Created by crossing Robusta 5 with Ottawa 3, this selection produces a tree averaging 45 to 50 percent of standard size. It has shown good compatibility and is well anchored. It is more tolerant of latent viruses than Geneva 935. G890 is increasing in popularity. Hardy to Zone 3b.

G935 (Geneva 935): This selection is a cross between Ottawa 3 and Robusta 5. (Note that even though the parents are the same as G890, every cross will produce a different result.) G935 is in the same size range as M26 (30 to 40 percent of standard), is very precocious and productive, and induces good fruit size. It is highly resistant to fireblight and collar rot and is easy to propagate in the stool bed. This is a promising rootstock, though it is not considered compatible with Royal Red Honeycrisp (a dark-red sport of Honeycrisp), Pazazz and certain other cultivars. Virus-free wood should be used if possible. Hardy to Zone 3b.

Ottawa 3: This rootstock was developed by Agriculture Canada at its Ottawa Research Station. It is a cross between M9 and Robin Flowering Crab. It is slightly larger than M9 and should be staked. It is an efficient and productive rootstock that is very hardy except in the coldest of circumstances. It is slightly susceptible to fireblight and woolly aphid. Ottawa 3 is challenging to propagate and is now hard to find. Hardy to Zone 3b.

P2 (Podkladki 2): This is a Polish dwarfing rootstock similar to M9 (40 percent of standard). The "P" refers to *podkladki*, the Polish word for rootstock. It is reportedly freestanding, but most growers do stake these trees. At the end of the season, the trees shut down earlier than most, and they begin growth slightly later in spring — indicators of good hardiness. Hardy to Zone 3a.

P22 (Podkladki 22): This is another Polish rootstock that produces very dwarf trees, approximately 20 percent of standard. P22 is a cross between M9 and Antonovka, a hardy Russian cultivar. It is resistant to most diseases, though susceptible to fireblight and woolly aphid. This rootstock requires staking. Hardy to Zone 3a.

Sprout Free: A vigorous selection that is very resistant to suckering. The rootstock is used mostly for propagating flowering crabapples. Hardy to Zone 3.

V1 (Vineland 1): This rootstock, developed by Guelph University at its Vineland Research Station in Vineland, Ontario, has shown excellent hardiness. It produces a tree approximately 50 percent of standard and has shown good fireblight resistance as well as better yield performance and efficiency than the standard rootstock in this size range. Hardy to Zone 3a.

V2 (Vineland 2): This is another Vineland introduction that has shown good hardiness and yield efficiency. It produces a tree approximately 50 percent of standard size. It has good fireblight resistance. Hardy to Zone 3a.

V3 (Vineland 3): This Vineland introduction has been getting favorable reviews from those trying it in Zones 2 to 3. It is very dwarfing, producing a tree approximately 30 percent of standard, similar or smaller than M9, the standard of this size range. V3 should be staked. It is considered resistant to fireblight. Hardy to Zone 3a or perhaps 2b.

An interstem is the inclusion of a dwarfing section of wood fitted between the rootstock and the bud of the desired cultivar.

Interstem Trees

It is possible to combine the attributes of a hardy, well-anchored seedling or clonal rootstock with the dwarfing abilities of a hardy, dwarfing clonal rootstock to produce a productive, small tree that is freestanding and hardy throughout. Such a tree is called an interstem.

A section of dwarfing rootstock is placed between the hardy, well-anchored rootstock and the fruiting cultivar. This can be accomplished in a two-stage process by grafting or budding the interstem piece onto the rootstock and then grafting the cultivar onto the interstem the following year. It is also possible to graft a bud of the fruiting cultivar onto a section of interstem and then directly graft this section onto the rootstock, thus eliminating a year from the process.

The interstem tree has the advantage of a tough, well-anchored root system for support and winter hardiness combined with the dwarfing abilities of the interstem section, which influences the productivity of the scion.

The length of the interstem section is critical to the amount of dwarfing imparted to the tree. Research indicates that 6 in. (15 cm) of interstem produces the maximum amount of dwarfing. Shorter interstems will produce progressively larger trees. Longer interstem sections, however, do not enhance the dwarfing effect.

While this process is somewhat more expensive and time consuming, it produces a tree that offers many advantages to the northern grower. Those who are considering interstem trees should contact nurseries to custom graft the combinations of rootstock, dwarf interstem and cultivar they desire.

Grafting Tools and Materials

Before entertaining a notion to graft there are tools and materials that must be acquired to complete the task successfully. In addition to the appropriate rootstocks, which we discussed earlier, here are the most important.

Scion Wood

Scion wood is the wood you remove from a mother tree of the cultivar you want to reproduce. When spring grafting, the scions should be gathered in late winter before there is any swelling of the buds. In our Zone 4 site this is in late March. It is best to collect scion wood when it has been above freezing for 48 hours. Scion wood should be gathered from wood that grew the previous season (one-year-old wood). This new wood usually contains only vegetative buds, which will form shoots. The older wood may have buds that are preconditioned to grow spur wood for future flower production and are poor choices for grafting.

Buds from the very tip of the scions, or from the very base are usually not used. The tip buds are those most likely to have suffered winter injury, and the terminal bud will open before any buds below it. The basal buds are more dormant and can take much longer to initiate growth. The central section of the new growth has the most ideal buds for grafting.

In older trees that have not been pruned recently, it may be difficult to find vigorous new wood. Usually the best wood will be found at the top of the tree on the south side. If the tree is old and tall, you'll need to find a safe way to gather the scions.

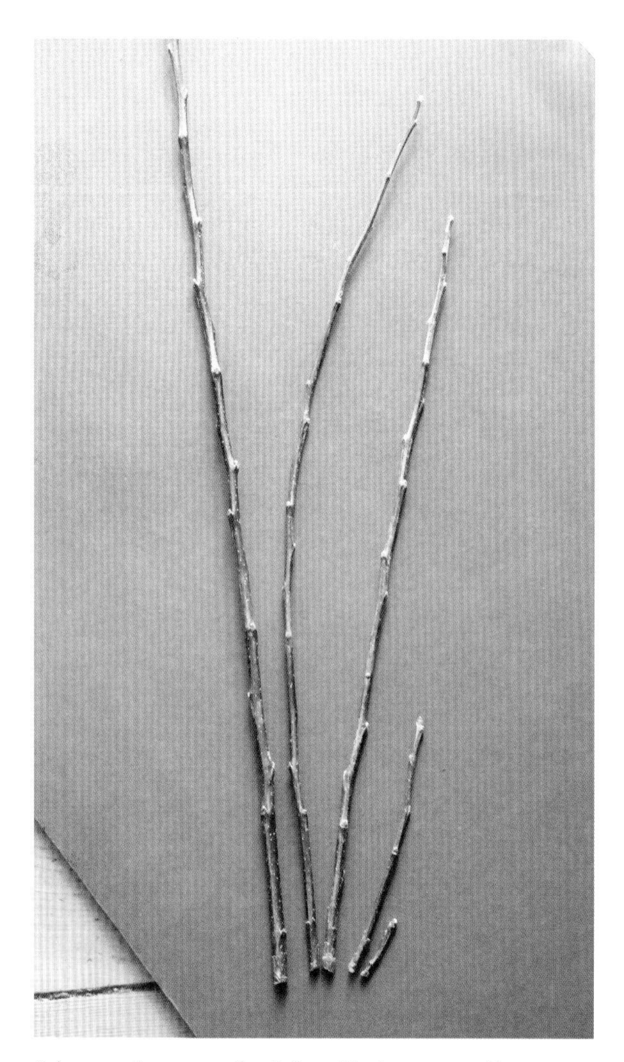

Scion wood can come in all sizes. The larger wood is the most desirable, as long as the rootstocks are the appropriate size.

Commercial nurseries generally maintain stock blocks of cultivars. These plants are hedged periodically to provide vigorous, large-diameter scion wood. This enables the grafters to use larger-diameter rootstocks. Larger scion wood will generally produce more vigorous growth in the field.

Once collected, place the scions in a sealed polyethylene bag, label it and keep near freezing,

28.4°F to 35.6°F (–2°C to 2°C). A slightly moistened paper towel around the bases can be used, though it is not usually necessary. Excess moisture can rot the scions and should be avoided.

If you are using a refrigerator to store the scion wood, there should not be any fruit in it. Ripening fruit gives off ethylene gas, which can be fatal to buds over a very short period of time. If you do not have a refrigerator available, you can bury the bags under snow drifts on the north side of a building or even in the soil, though you will need to ensure it's in a place where the sun will not heat it until you are ready to graft.

The Knife

To perform grafting you will need a very sharp, thin-bladed knife, preferably one with a blade that is beveled on one side, in much the same manner as a chisel. When pulling through the wood, a single-bevel grafting knife will leave a flat surface. When aligned, two flat surfaces create good surface-to-surface contact. If the knife has a bevel on both faces, the knife will have a tendency to form a cupped surface as it is drawn through the wood. Cupped surfaces can leave air space between the cuts when aligned. Even if these are forced together, tension will remain and may eventually cause the graft union to fail.

If you are going to perform any number of grafts it will make the job far easier if you have a knife made for the purpose. If you decide not to use a grafting knife, use a knife with a very thin blade, such as a utility knife. If you are skilled in honing, you can create a single-bevel blade using a sharpening wheel or stone.

A grafting knife.

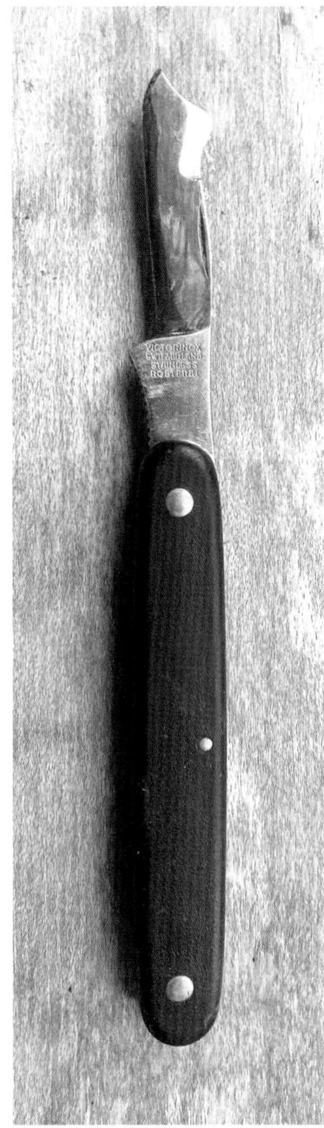

A budding knife.

When budding, it also pays to use a true budding knife, since budding is a slightly different process. A budding knife will be beveled on both sides and will have a rounded "horn" on the upper side of the knife near the tip. This horn is used to pry apart the bark corners after the initial cuts are made, so that the bud can be inserted.

Keeping your knives razor sharp is one of the secrets to successful grafting and budding. If you purchase a new knife, be sure to maintain the angle of the blade when honing. There is a tendency to increase the angle to "catch" the edge. It is far better to keep the stone on the angle of the bevel and hone down until you reach the edge. When doing the backside of a grafting knife, be sure to keep the blade flat on the stone.

Although initial sharpening can be done with a carborundum (silicon carbide) stone, the final sharpening should be done with a very fine stone, such as an Arkansas stone, which is made from naturally occurring novaculite, or a very fine water stone. When final honing is done, the knife can be stropped on a piece of leather.

The amount of sharpening will depend on the amount of grafting you do and the cleanliness of the scions and rootstocks you are using. It is critical that the rootstock in particular be wiped with a damp cloth to remove any grit on the surface. Such grit will rapidly dull the knife. The key factor is the quality of the cuts. The knife should slice through the wood easily and leave sharp edges and a perfectly flat surface. Once the edges start shredding or the pulling becomes harder, it is time to sharpen.

Grafting Machines

Over the years several different types of grafting machines have been developed to speed up the process of grafting. These machines, if well designed and kept sharp, can be useful in large operations. For individuals or small businesses, the cost/benefit of such machines is more problematic.

An omega grafting tool.

This author believes that though there may be a place for such mechanization in certain operations, these machines cannot create the precision that can be achieved by a talented grafter with a sharp knife.

There is a handheld device called an omega grafter that creates a key-shaped cut that will work well for an unskilled grafter, but again a knife will create a finer cut.

Disinfectants

The grafting area, tools and materials should be kept orderly and clean. Although perhaps not as critical as in animal surgery, the plant surgery we call grafting can fail if harmful materials are allowed into the graft union. There may be bacterial or fungal cells on either scion wood or rootstock adhering to the surfaces that

Traditional grafting wax contains tree rosin and beeswax. Parafilm® is used to wrap the graft before the wax is applied.

can be spread by the knife. To combat infection, it is highly advisable to disinfect your knife regularly and always between cultivars. The most commonly used disinfectant is rubbing alcohol, which can be bought in any pharmacy.

Keep a small bottle of alcohol with enough liquid that the entire blade of the knife can be dipped. Then wipe dry with a clean cloth or paper towel. Alcohol leaves no residue and does an excellent job of sterilizing surfaces.

Tying Materials

Grafts need to be tied together to keep the graft union aligned and to exclude air. In the past materials like raffia or waxed twine were used to tie the union. These can still be used, but the disadvantage to these is that, if not removed, they will constrict the growing stem. The most commonly used material today is Parafilm®, a stretchy, waxy plastic that adheres to itself. It is used in laboratories and hospitals to create airtight coverings on beakers and petri dishes, among other uses. Because it is self-adhesive, there is no tying required. The bud will grow through the Parafilm®, and therefore you do not need to cut the tie later.

When budding, the bud is often secured using narrow rubber budding strips. These strips press gently on the front of the opening to ensure it stays tightly closed until the bud union has healed. Polyethylene strips are sometimes used as well.

Wax

Wax is an ideal material to ensure no air penetrates to the graft union. There are several commercial grafting waxes available. Most come as a bar that can be heated or molded with the heat of your hands. The traditional grafting wax that has been used for centuries combines beeswax and tree rosin. Mixtures vary, but a rule of thumb is to melt together three parts beeswax to one part tree rosin. The wax helps keep the union flexible in cool weather, while the rosin keeps it stiff in warm weather.

When grafting indoors you can use a controlled temperature vessel, such as a slow cooker, to keep the grafting wax melted but not overly hot. If the wax is too hot it can damage the tree tissues. Once tied, the newly made graft can be quick dipped into the wax.

Working outside is trickier as the wax will harden as soon as it leaves the container. Most outside grafters use a large can with a handle. Inside the can is a shorter can that is suspended above the base of the exterior can. The base of the exterior can has air holes to allow a heat source, such as a jellied alcohol (e.g., Sterno®), to burn under the suspended can that holds the wax. You can also use a propane torch to occasionally heat a pot of wax, but it is difficult to keep the temperature at the proper level.

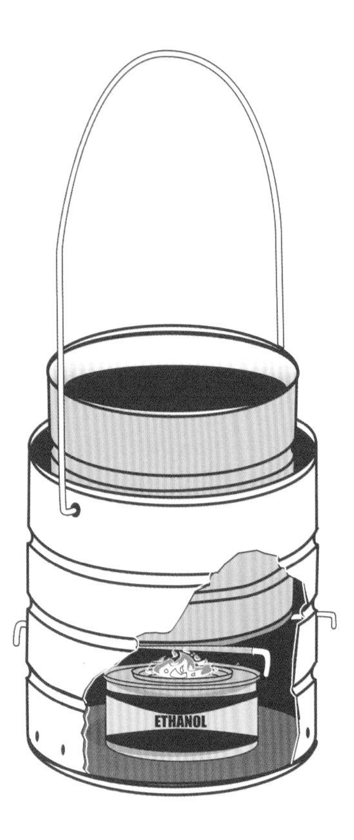

For outside work the wax is usually applied with a small brush. Because the cool air will harden the wax on a brush very quickly, it is necessary to dip and redip constantly. The brush should have a device to hold the brush end in the wax, while keeping the handle above and readily accessible. You should definitely wear thin gloves, as it is a sticky operation. After this description you might want to use moldable grafting wax that can be heated in your hands. This is fine as long as air does not get into the graft.

Organizing the Grafting Operation

If you are doing any amount of grafting, especially if several grafters are involved, it is vital to keep the grafting area organized. As the grafter works, there can be sections of both rootstock and scion wood lying on the table. It is easy to have mix-ups in such situations. To a novice, or even an expert, the scion wood of different cultivars can look nearly identical. Try to remove any rootstock discards from the working table as you work. Completely clean up the table after each cultivar is finished and before a new one is started.

When working outside, either grafting or budding, the same rules should apply. Strict adherence to the organization of scion wood and bud wood should be your mantra.

Setup for grafting should be kept organized so no mix-ups occur.

Grafting and Budding Methods

There are many methods used to graft trees. Though each is a process designed to accomplish the goal of intimately securing together the cambiums of the rootstock and the scion, the mechanical tasks and seasonal timing of the procedures differ. Listed below are some of the more common forms of grafting and budding. There are many variations on these, but most propagators of fruit trees rely on the methods listed here.

Whip and Tongue Grafting

This is the most commonly used grafting method. It is the method of choice when grafting scion wood onto bare-root dormant rootstocks in early spring, a process often referred to as bench grafting. The whip and tongue method can also be used outside to top-work young or mature trees that are in the ground.

Although rootstocks used in spring grafting are usually dormant, it is acceptable, some would say preferable, that the rootstocks are just beginning to come out of dormancy. Such rootstocks will begin cell division more rapidly than completely dormant rootstocks. Be sure to keep the roots moist at all times, but never allow them to sit in water for more than a short while. Containers with moist sawdust work

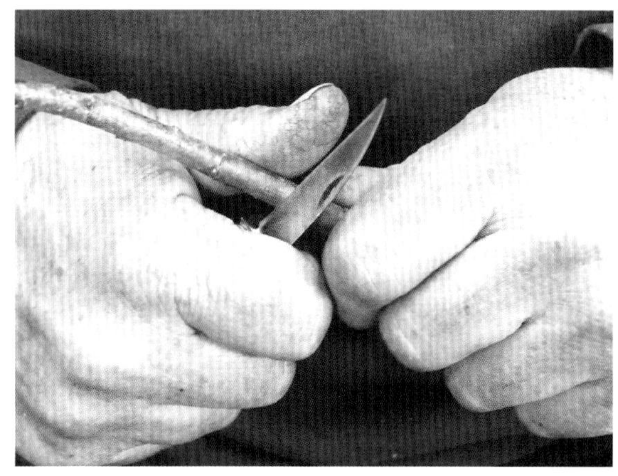

Make the first cut quick and sure with fingers and arms braced for control.

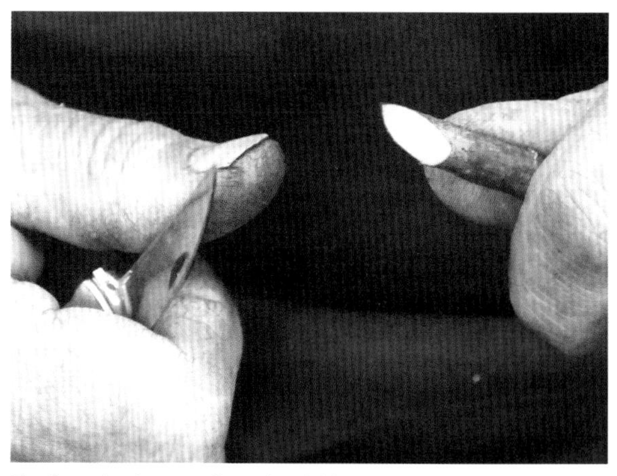

The length of cut surface should be two to three times the diameter of the wood.

Starting at the center, make the second cut as thin a flap as possible.

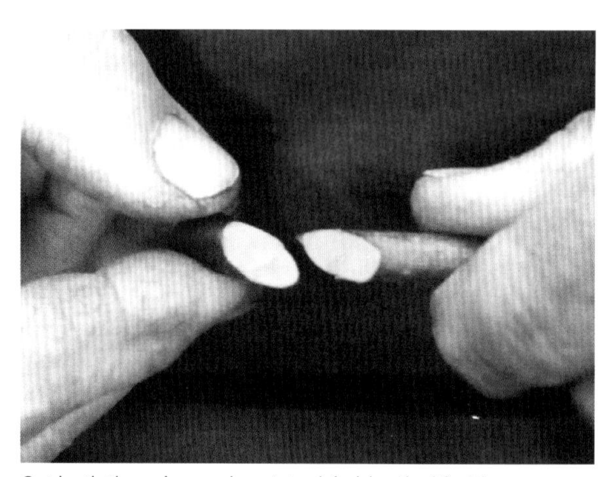

Cut both the scion and rootstock in identical fashion.

The cambiums of at least one side should be perfectly aligned. With luck, all edges of the graft will have good contact. This will result in a stronger union.

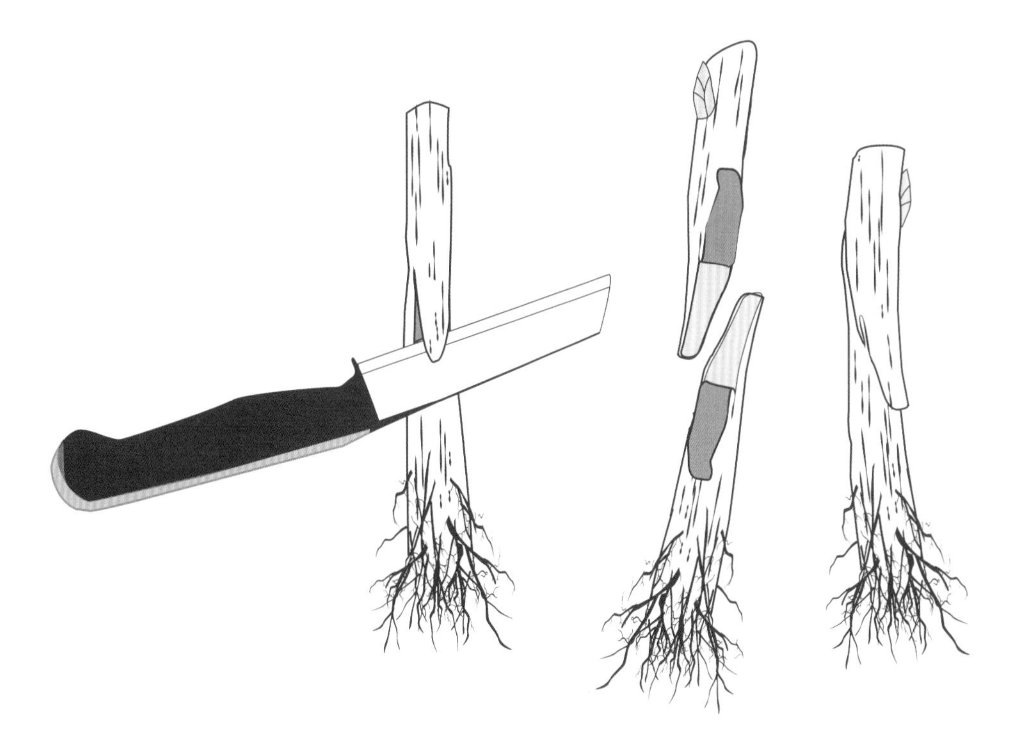

well as you can keep the roots moist but not overly wet.

The whip and tongue technique involves making an identical angled cut on both the rootstock and scion with a sharp grafting knife. While most texts will show a rather long cut, it is far better to keep the length of the cut only two to three times the diameter of the stock. This avoids the "flaps" that are created when the cuts are too long. Flaps can prevent the two ends of the cuts from healing properly and may cause burr knots to form.

Make the initial cut with a quick, sure motion, rather than "working" your way through the wood. A quick motion done properly will result in a smooth flat surface. Depending on the grafter, the required motion is to either jerk the knife through the wood toward you or quickly pull the wood along the blade. For some it is a balance of these two motions.

In this method of grafting, the knife is drawn toward you — something the safety conscious will admonish you for doing. To work safely it is vital to control the cuts by positioning the fingers on and around the knife and by bracing your arms against the sides of your body. An uncontrolled cut can lead to serious injuries. In the case of grafting, a dull knife will increase the probability of injury. A sharp knife should glide easily through the wood and not require the buildup of pressure that is released suddenly when a dull knife finally makes the cut.

Make a second cut at the center of the first cut on both the scion and rootstock. This second cut is made as shallowly as possible. It locks the two sections together, making it easier to align and wrap the graft. If the second cut is not shallow the surfaces will spread apart when pushed together. This cut generally takes longer to master.

The second cut is made by catching the edge of the knife in the wood, then lowering

When the scion is smaller than the rootstock, place one side in alignment.

The graft should be wrapped entirely in Parafilm® before waxing. The bud will push through both film and wax.

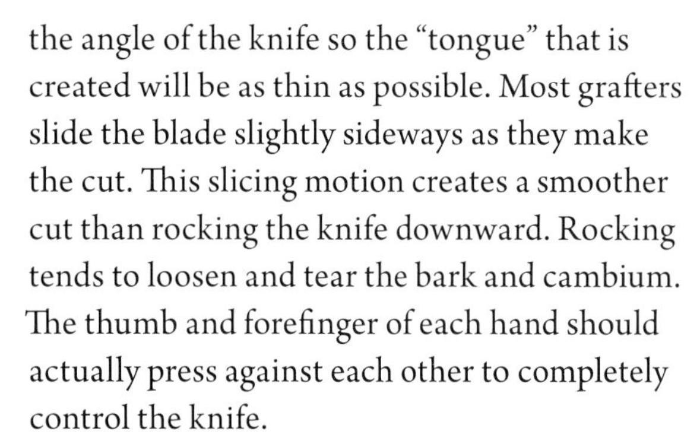

the angle of the knife so the "tongue" that is created will be as thin as possible. Most grafters slide the blade slightly sideways as they make the cut. This slicing motion creates a smoother cut than rocking the knife downward. Rocking tends to loosen and tear the bark and cambium. The thumb and forefinger of each hand should actually press against each other to completely control the knife.

Carefully matching the diameters of both scion and rootstock stem is critical if the graft is to be successful on both sides of the union. A good match will allow the cambiums to be aligned around the entire circumference of the union.

It is sometimes the case that the only scion wood available is smaller than the rootstock. In such cases you must be sure at least one side of the graft is perfectly aligned. Do not center the scion piece or you will have no cambial contact and the graft will fail.

While most texts show a scion piece with several buds being used for grafting, it is more efficient to use a single bud. If a multi-bud scion is used, you must return to the field to rub off all but one bud to establish a single dominant leader. This process is time consuming and carries the danger of breaking the delicate union in the process. By using only one bud you need only rub off suckers coming from the rootstock.

Once you have fit the scion bud on the rootstock, cut the scion with your pruning shears slightly above the bud closest to the rootstock at a slight angle away from the bud. If there is a second bud below the bud you selected, simply remove it by placing the knife at the base of this bud, pressing gently and slicing it off.

Once the graft is made and aligned, place a strip of Parafilm® — approximately 1 in. (2.5 cm) wide and 4 in. (10 cm) long — to the base of the graft. Gently holding the base of the union so that it does not move, wind the film around the graft using enough pressure to slightly stretch the film but not so much as to undo the alignment. The film can be stretched over the top of the scion piece and then wound back down over the union. When done, simply press the film against the rootstock with your index finger and it will twist off. Soon after the graft is completed, dip or brush over the entire union in grafting wax.

If you are grafting onto seedling rootstocks it is acceptable to graft quite low on the rootstock. If you are grafting onto dwarfing rootstocks it is advisable to leave 6 in. (15 cm) between the crown of the rootstock and the graft. This will ensure that the scion, which is most often more vigorous than the rootstock, will not be able to root. Such roots would soon negate any dwarfing influence of the rootstock.

Once you have finished, the grafts should be temporarily placed in containers with a damp material, such as sawdust or fine bark, that is deep enough to cover the entire root system. Be sure the containers have drainage

New shoots emerging from buds. Dwarfing roots should be grafted 6 in. (15 cm) above the soil to prevent the scion from rooting in the future.

holes. The roots need to be kept moist but never waterlogged.

Keep the grafts out of the wind and at a nominal temperature of 50°F (10°C). This temperature will allow the grafts to heal before the scion buds begin to expand. Once the buds form green tips, the grafts, even if still in the containers, should be placed in the sun. They can be left outdoors as long as temperatures do not dip below freezing. If it is still too cold, bring the grafts in at night. If you are grafting in midwinter leave the grafts to heal at 50°F (10°C) for two weeks, then place them in a cold room at 36°F (2°C) until they can be planted.

You should be ever mindful of the delicacy of the graft unions. Be sure they are never allowed to push against each other. Keep the grafts out of high-traffic areas, where an errant arm or foot could cause irreparable damage.

The finished grafts can be planted once temperatures are unlikely to dip below 28°F (–2°C). The pH of the soil should be adjusted to between 6.5 and 7 for optimal growth. Be sure the roots are spread evenly in the planting hole or planting row. Incorporating bone meal, blood meal and/or fish meal near the roots will give the new graft sufficient nutrients for good growth. Using mycorrhizal inoculants is also an advantage.

In commercial plantings, the use of the bacterium *Agrobacterium radiobacter* is highly recommended. There are several commercial formulations of this bacterium. It inhibits the growth of the crown gall bacterium (*Agrobacterium tumefaciens*) on the roots. Crown gall bacteria form irregular galls on the roots and can reduce the vigor of the tree in the future. *Agrobacterium radiobacter* is very specific and poses no danger to other life-forms. This is a far better solution to the problem than the soil sterilization that was formerly practiced.

Agrobacterium radiobacter comes in a powder form that should be kept refrigerated. Just before use, mix the bacteria with a small amount of water to form a paste and then mix the paste with clean cool water in a container such as a clean plastic garbage can. The container should not allow light through the sides and a cover should be kept on until use as sunlight will kill the bacteria. Simply dip the roots into the water just before planting. There will be enough bacteria adhering to the surfaces of the roots to protect them against the crown gall bacteria in the soil.

Whip and tongue grafting can also be used to graft scions onto smaller branches of existing trees. However, there is a more common method of grafting branches of an existing tree to another cultivar that is discussed in the cleft grafting section.

Cleft Grafting

Cleft grafting is quite possibly the oldest method of grafting. Today it is most often used to graft branches of an existing tree to a different cultivar. This method is performed in early spring, just as the buds are swelling; however, it is important that the scions you intend to use were gathered earlier when the tree was completely dormant and have been kept refrigerated.

This technique involves cutting off a branch or stem, usually with a diameter in the 1 to 2 in. (2.5 to 5 cm) range, though larger branches can

Newly cleft grafted scions on a "mother" tree chosen for the purpose.

New growth from the fully healed scions.

Close-up of a healed cleft graft. Note the triangular shape of the scion base.

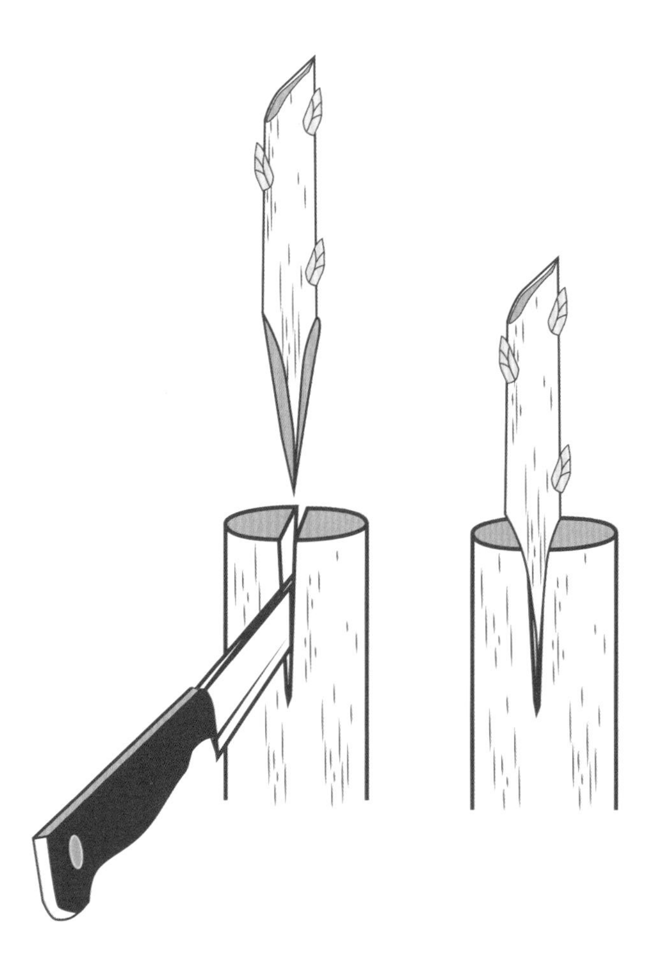

be used. A sharp knife makes a good tool for splitting the end of the branch where the scions will be inserted. Place the knife in the center of the cut and tap the dull edge lightly with a mallet or hammer, being careful the split does not extend more than a few inches. Use a flat head screwdriver or similar tool inserted into the center of the split. By twisting gently you can open the split to receive the scions on one or both sides of the split.

The scions are prepared by slicing each side of the scion to form a long wedge at its base. Most grafters use a scion containing three to four buds. A larger scion in this case usually means better vigor. Examine the wedge to determine which side of the wedge is the widest and place this side on the outside of the branch in the split. The outsides of the barks of both stock and scion should be contiguous. The wedge must be aligned carefully so that the maximum amount of cambial contact is achieved. If the branch is wide enough, a second wedge is placed on the other side. Once the scions are aligned, the tool holding the split can be released slowly.

Usually little or no binding is required to hold the scions in place, as the spring back pressure of the split branch will usually keep the scions from moving. If you feel the pressure is insufficient, wind masking tape tightly around the site. Masking tape will break down as the branch expands. Electrical tape works well, but this tape must be sliced or removed later or it will strangle the branch. Once completed, coat the entire surface of the operation in wax to prevent air from entering the split or the sides where the wedges have been placed.

If both grafts are successful, it is best to keep the weaker-growing side pruned back for several years. The stronger scion will form the new branch. The scion that is kept pruned will continue to heal the cut, but will be kept small and can be sliced off when callus tissue has completely covered the cut.

This graft is relatively easy to perform and is often used by new grafters, but it has drawbacks. The fact that only one side of the scion unites and that the split can remain open for several years can create a weaker union and may give entrance to fungal diseases.

This technique does allow an older tree to be top-worked to a different cultivar. Growers who want to switch to another cultivar can use the cleft graft on wider-diameter branches to create a productive tree in a short time. Because the scions are grafted into the fruiting area of a tree, they begin producing within two years, allowing crops of the new apple cultivar to be harvested much faster than if you started with young nursery trees.

When grafting on older trees it is wise to spread out the top-working of the tree over two years so that the tree will have enough leaf area left to produce sufficient food for good growth. Another option is to graft the branches you wish, but leave several smaller branches throughout the tree to draw up sap to feed the new growth and the roots. This will speed the transition and prevent the starvation and death of the root system. When the new branches are growing vigorously the smaller, sap-drawing branches can be removed.

T-Budding

The T-bud has, until recently, been the most frequently used method to produce apple trees. However, chip budding (see pages 133–134) is quickly overtaking the T-bud's position for summer budding.

T-budding is used on rootstocks that are actively growing in the field. In cooler regions most budding is done in August. The rootstocks should be a minimum of 0.4 in. (1 cm) in diameter. Growers should irrigate the rootstocks if the soil is not moist. It is critical that the cambium be able to "slip off" the internal wood, for this is where the buds will be placed. If the rootstocks are not well hydrated the cambium will adhere to the wood, and the chances of success will be minimal.

This technique can be trying on the budder's back. Most often the budder is leaning over the rootstock to make an upright T-shaped cut at the desired height. Using the horn of the budding knife, both corners of the T are opened and a bud is inserted between the cambium and internal wood. As mentioned, the interface between cambium and wood should be "slippery."

The buds are cut from scions gathered from the current season's growth. It is important that the scion material be actively growing. Immediately after harvesting, the leaves should

Gather bud sticks from new growth. Retain the petioles for handles.

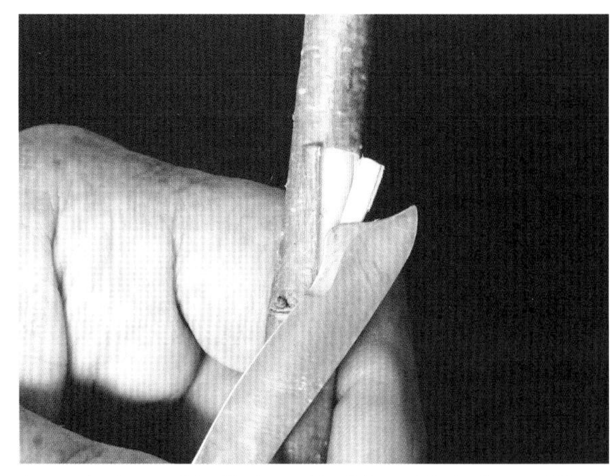

Open the flaps of the T-shaped incision with the horn of the budding knife.

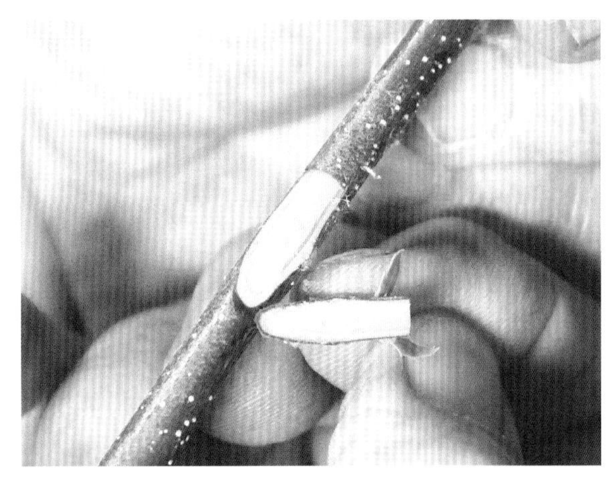

Slice the bud from the scion stick and remove the interior wood.

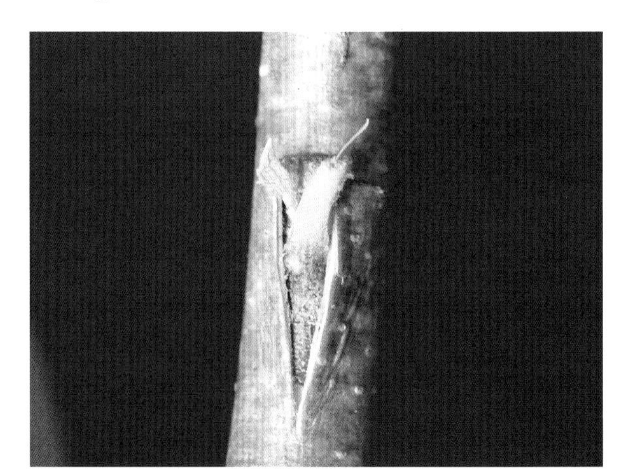

Insert the bud.

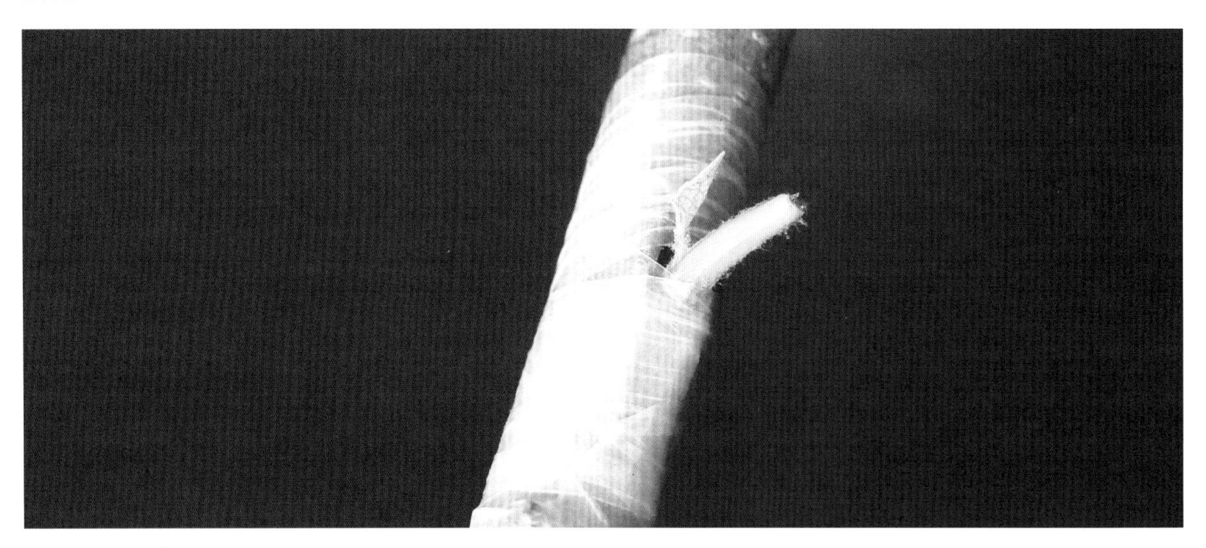

Using Parafilm®, wrap the site except for the actual bud. In spring cut the rootstock just above the bud and train the new shoot. Remove any suckers from the rootstock as soon as they appear.

be removed, taking care to leave a small section of the petiole next to the bud. This will serve as a handle to help insert the bud.

The buds are removed from the scion by slicing shallowly upward from just below the base of the bud, then making a shallow slice just above the bud, a slight rocking motion that cuts through only the bark and cambium, stopping at the wood. Grasp the petiole and with forefinger and thumb on each side of the bud, press the bud gently and slide sideways, removing the entire bud from the interior wood. This shield-shaped bud section is then slid into the T-cut and wrapped.

Some budders wind narrow strips of polyethylene around the entire area while leaving the bud itself exposed. Many today wrap the entire area in Parafilm®. The bud will grow through the film and will not require removal later, as will the polyethylene. An older method involves wrapping thin budding rubbers around the site, again leaving the bud itself exposed.

Under the bark, cells multiply until the cambium of the rootstock and the cambium of the bud shield unite. This must happen before severe frosts occur. Early the following spring the rootstocks are cut off at an angle slanting away from, and just above, the bud. If all has gone well, growth will begin.

This method offers some advantages. Using rootstocks that are established in the ground gives great vigor to the bud and can produce tall straight whips in the first year. Budding can be accomplished in late summer when most nurseries are less busy. The method is fast for those who are trained, and production rates can be much higher than with most other methods. Some budders have been known to bud 4,000 rootstocks in a single day with two tyers working behind them.

Budding does have some disadvantages, though. Because the cambiums of the bud and rootstock are not initially in tight contact, the tissue that forms between them is relatively weak for the fall and early spring growth period. In colder climates the scion and rootstock might not unite before heavy frosts occur at night, which can cause serious crop failure. This can also result in the new bud being snapped off its rootstock in the wind at the start of the growing season. Although the buds must be at the proper stage of growth, which is usually early August in cooler areas, budding as early as possible may lessen this problem.

Chip Budding

The chip bud is used extensively today and is rapidly gaining preeminence. In this form of budding the rootstock is prepared by making a 45-degree cut downward to a depth approximately one-third the diameter of the rootstock. The second cut starts about 0.75 in. (19 mm) above the first cut. The knife is slid down to meet the base of the first cut. After the second cut is made, the chip is grasped between the knife blade and thumb, removed and discarded. The surface left after the chip is removed is an arc shape referred to as the "church window."

The bud sticks or scions are prepared in much the same way as for a T-bud, but the buds are removed by making cuts beneath and

Cut the rootstock downward at a 45-degree angle, then make a second cut down to the base of the first cut.

Cut a matching piece from the scion stick and insert it snugly. Bind with Parafilm®.

behind them that exactly duplicate the shape of the chip taken out of the rootstock. The bud is held between the blade and thumb so that cut surfaces do not come into contact with the fingers. The bud is then slid into place and gently pressed into the notch at the base.

Once in place, the bud is wrapped with either polyethylene or Parafilm®. Because this is a true graft, the cambiums are in direct contact, and healing takes place more rapidly and a stronger bond results.

The chip bud method can be performed in spring or fall. Spring budding can be used on rootstocks in the ground or on bare-root rootstocks at the grafting table. Late summer chip budding is usually performed in August.

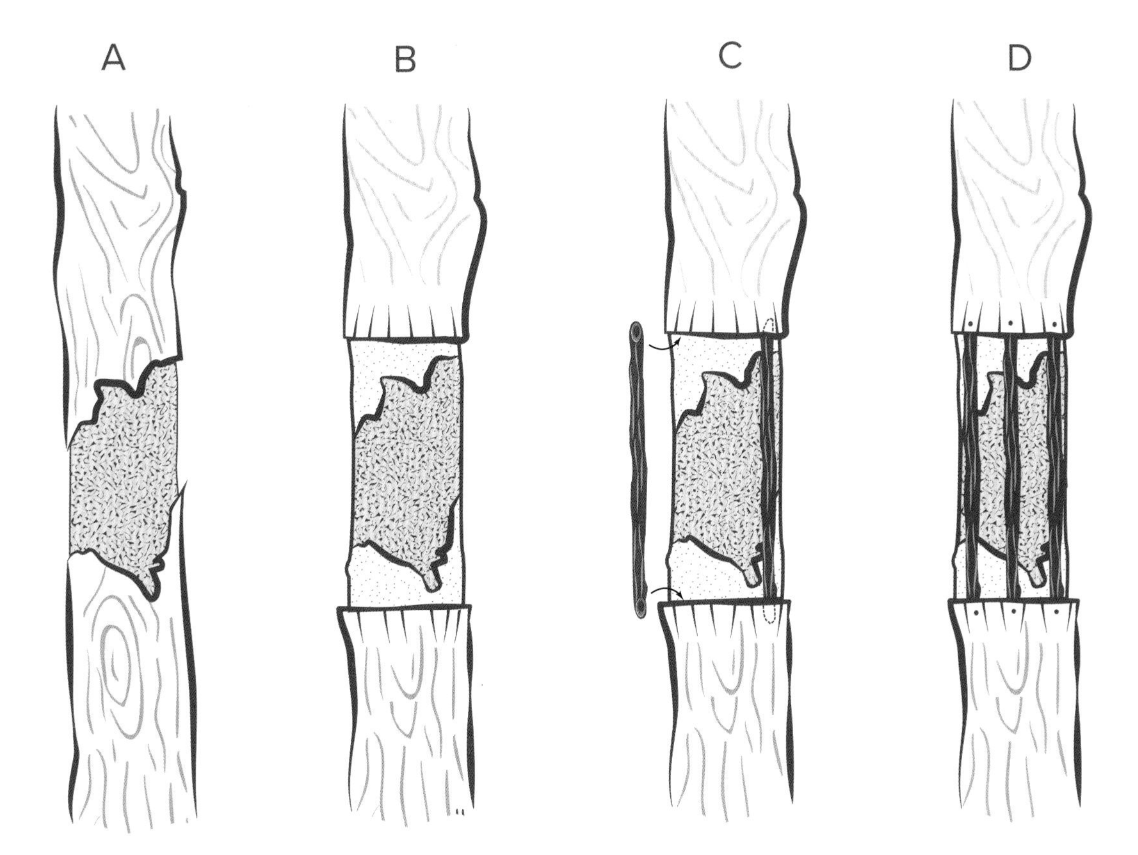

Bridge Grafting

Bridge grafting is a technique employed to save trees that have been injured by mice, voles, rabbits or lawn-mowing equipment. It is a challenging technique and less successful than other methods, nonetheless it does offer hope of saving valuable trees.

Bridge grafting must be done as soon as possible after the damage (A) occurs. This is usually in early spring, when the grower first examines their trees.

The upper and lower edges of the wound are cut horizontally back to where the bark adheres to the wood (B). Several scions, their number determined by the size of the wound, are prepared to be slightly longer than the length of the wound. Cut a tapered wedge on both ends of the scions along the same axis. Make two shallow vertical slices to the top and bottom of the wound, where each scion is to be placed, insert the wedges under the bark between these cuts (C) and drive a brad in each end to hold the scion in place (D). The more scions you can fit, the better, though it is usually practical to place them no closer than 1 to 2 in. (2.5 to 5 cm) apart. When completed, apply wax to the graft sites to prevent drying.

Depending on circumstances, this method may work to save a tree. Often, not all the bridges will survive, and the good done may be questionable. In all cases this graft, even if successful, will create a tree that is more subject to the stresses of wind and the intrusion of fungi and bacteria.

9 Cultivars for the North

The Cultivars

Alexander
Ambrosia
Antonovka
Ashmead's Kernel
Bailey Sweet
Ben Davis
Bethel
Black Oxford
Blue Pearmain
Bonkers (NY 73334-35)
Bottle Greening
Bramley
Carroll
Chestnut Crab
Cortland
Cosmic Crisp
Cox's Orange Pippin
Crimson Crisp
Delicious (Red Delicious)
Dolgo
Duchess (Duchess of Oldenburg)
Dudley (Dudley Winter)
Empire
Enterprise
Fameuse (Snow Apple)
Fireside
Freedom
Frostbite
Gala
Ginger Gold
Golden Delicious
Golden Russet
Goodland
Granite Beauty
Greensleeves
Haralson
Honeycrisp
Honeygold
Hudson's Golden Gem
Hyslop Crab
Keepsake
Liberty
Lobo
Lodi
Macoun
Mann
McIntosh
Milwaukee
New Brunswicker
Norland
Northern Spy
Northwestern Greening
Novamac
Parkland
Patten Greening
Patterson
Paulared
Pewaukee
Pomme Grise
Priscilla
Pristine
Pumpkin Sweet (Pound Sweet)
Red Astrachan
Red-Fleshed Crab (Hansen's Red Flesh)
Redfree
Rhode Island Greening
Roxbury Russet
Sandow
Seek-No-Further (Westfield Seek-No-Further)
Silken
SnowSweet
Spartan
Suncrisp
SweeTango
Sweet Bough
Sweet Sixteen
Tangowine
Tolman Sweet
Viking
Wealthy
Wickson Crab
William's Pride
Wolf River
Yellow Bellflower
Yellow Transparent
Zestar!

Cider Apples

Banane Amère
Bilodeau
Brown's Apple
Bulmer's Norman
Dabinett
Douce de Charlevoix
Geneva Crab
Harrison
Kingston Black
Muscadet de Dieppe
Yarlington Mill

◂ Liberty apples ripening on the tree.

There are many cultivars of crabapple with pink and red flowers. These are derived from the tongue-twisting Russian species *Malus Niedzwetzkyana*.

There are thousands of named apples. Sorting through the immense number of cultivars and deciding which to highlight is a daunting challenge, especially for someone who wants to list them all. A great percentage of apples are not suitable where winter temperatures are severe, and some apples, though hardy, are difficult to grow or are unproductive. The following selections have been chosen for characteristics such as hardiness, outstanding flavor, culinary excellence, disease resistance and/or historical importance. Some of these apples are exceptional, but I believe all are interesting and useful. However, many may find this list wanting; every list is limited.

As the world grew more interconnected, the number of known apple species expanded. Many apple species are small fruited and have been labeled "crabapples." An arbitrary size limit of 2 in. (5 cm) has been established to designate an apple as a crabapple. Today there are an immense number of hardy and notable crabapples, and their numbers increase every year. The desire to breed new and improved ornamental crabapples has created new colors and forms that add delightful splashes of color to the landscape.

I have decided not to include crabapples in this book except for a few of the better-known edible crabs. For those who wish to delve into the fascinating world of crabapples, I highly suggest you find *Flowering Crabapples* (1995) by Father John L. Fiala. This book is the most complete compendium of crabapple species and cultivars that has been written to date. It is well worth searching out.

A Note on Taste

Descriptions of an apple's taste can be objective only when judging a cultivar's level of acidity, sweetness or tannins. These qualities can be measured analytically in a laboratory. As well, the tongue acts as an accurate measure of these basics. For example, it is easy to tell an acidic apple from a more neutral or sweet apple.

Nuances of flavor, however, are not quantifiable for most of us. Perhaps scientists can measure various amounts of volatiles and other molecules that contribute to flavor, but describing the interaction of these molecules on our nose is exceedingly difficult because the vocabulary of smell as it relates to taste is inadequate.

The best way to judge the flavor of an apple is to take a bite.

How do we describe the subtle aromas picked up by the nose and palate that distinguish two apples? Most people could easily tell that a Cortland tastes different than a Northern Spy, but how do we express the subtleties that let us know that? A Cox's Orange Pippin has an acidity content of around 0.6 percent, which we can measure, but its flavor might be described as "aromatic" or "complex." What do these words really mean? What exactly is the aroma? How is it complex? Such language can only be a crude measure of aroma intensity or intricacy.

We often resort to comparisons with other foods. An apple may be said to have hints of pear, banana, raspberry, nuts or caramel. Distinguished palates might pick out such subtle aromas, but many will not. Those with an interest, however, can develop sophisticated abilities, becoming what one might call a "pommelier" — not unlike a sommelier.

What are we to do? How can we best describe the flavor of an apple? I guess we do the best we can and rely on the vocabulary of those pommeliers.

A Note About the Photos

Most of the tree photos were taken in our orchard, which was planted between 1981 and 1984. These trees are grafted onto Beautiful Arcade rootstocks, which form a semi-standard sized tree. The orchard was originally planted as a source of scion wood for grafting. It has been given minimal care through the years and has never been sprayed until the year these photos were taken, when it was sprayed with *Bacillus thuringiensis var. Kurstaki* (Btk) at petal drop to eliminate leafrollers and had weekly spot applications of GF-120 starting in early July to suppress apple maggot.

Because the trees have had only sporadic pruning through the years, the photos give a fair idea of the natural form of the cultivar.

Glossary of Apple Terms

Baking apple: Most often used to describe an apple that is outstanding when baked whole with the core removed and filled.

Basin: The depression at the blossom end of the apple.

Bloom: A waxy substance that covers the surface of some apples, often giving the fruit a hazy or smoky appearance. It is easily rubbed off.

Calyx: The collective name for the sepals of a flower, which are specialized leaves that are located at the outermost portion of the flower. The remains of the calyx can be found at the center of the basin.

Cider apple: An apple used for making alcoholic (hard) cider.

Cooking apple: An apple used for apple sauce or desserts such as pies, pastries and crumbles.

Dessert apple: An apple used for fresh eating and salads.

Lenticel: Respiration locations on the skin of the apple that appear as small dots. They can be white, dark or russeted in appearance, and they can be raised or depressed.

Pip: An apple seed.

Pippin: An apple seedling.

Russet: A reddish and/or tan area on the skin that resembles leather and is often rough to the touch.

Sport: A mutation occurring in a bud that produces a deviation from the normal type of growth or coloration of the fruit.

Gallery of Apple Shapes

Most apples will have profiles that fall into the shape categories shown below. Usually a single profile type will dominate, but there are often deviations among the apples of a cultivar, with some assuming a slightly different form. Thus, you may read a description such as "round to slightly conical." Be aware that insect activity and lack of pollination can distort the forms of apples, often drastically. When collecting apples for identification it is a good idea to pick three to five good specimens to get an average profile of the apple.

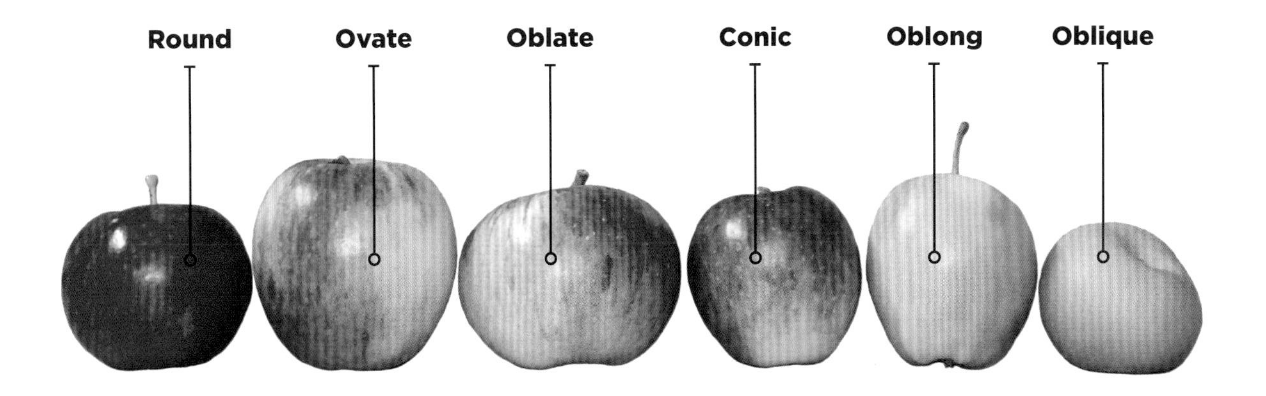

Alexander

Hardiness Zone 3
Introduced 1805
Origin Ukraine
Primary uses Cooking, baking

Alexander, one of a group of apples known as the Aport group, is believed to have originated in Ukraine. It was introduced into England in 1805, then later imported by the Massachusetts Horticultural Society in 1817. Its cultivation spread rapidly and was especially welcomed by growers in the colder areas of North America. Alexander was popular well into the early 20th century.

Though never esteemed as a dessert apple, it was grown extensively because of its superior hardiness, large size and culinary properties. Many have called it a "pretty" apple as well. In Atlantic Canada it is still one of the most commonly found apples in the oldest orchards. It is a proven survivor.

The apple is very large — no doubt a part of its appeal. It is round to conic with yellow-orange skin that becomes overlain with stripes and splashes of red where exposed to sun. The skin is waxy in feel and somewhat tough. The flesh is white to light yellow and coarse textured. Most would describe the flavor as fair in quality and the texture as akin to eating soft cardboard. It does, however, make excellent pies, and this was its main use.

The tree is large and upright, becoming somewhat pendulous with age. In more southerly climes it is reported to be somewhat prone to fireblight, but this has not proven to be a problem (so far) in the north. It is also slightly prone to apple scab and apple maggot.

Alexander seems to hold a place in the hearts of those who grew up with it. It was a pioneer among early hardy apples and its sheer tenacity has given it a longevity few apples can claim. Some think its quality is overstated, yet its popularity and fame cannot be disputed. ■

Alexanders are noted for their large size. When perfectly ripe, they are deep red.

Alexander forms a wide, spreading tree with age. It is an annual bearer and very hardy.

Developing Alexander apples show their characteristic round-conic form.

Ambrosia

Hardiness Zone 4
Introduced ca. 1990
Origin Cawston, British Columbia
Primary use Dessert

The best way to write about an apple is to be eating one while you are writing, so this is a captured moment. The apple in my hand came from a supermarket, where most apples end up these days. It is reminiscent of Golden Delicious in form, being conic to oblong. The yellow background is also similar, but it is the extent of the orange-red striping that differentiates it. The striping extends from the stem cavity to just pass the midpoint of the fruit. On the side where the apple was most exposed to sun, the stripes become a nearly solid blush, but not quite. Specimens grown under high light levels can be nearly all red. The basin has no reddishness at all and has prominent furrows similar to Golden Delicious, which makes sense, as it is thought to be one of its parents. The other is believed to be Starking Delicious.

The lenticels, or small dots, on the skin tend to be submerged (appear to be under the surface) at the basin, but on the stem end they tend to be ever so slightly raised and are white with a tiny dark dot in the center.

With little resistance, the fruit breaks off in chunks that are tender and juicy, but there is far more to the experience, as the aromatics make you feel like you are sipping a banana-pear-apple smoothie. Perhaps it is just my fancy, but I sense a few notes of vanilla in the background. It is definitely a sweet, fruity experience with just enough acidity to avoid blandness. The skin is barely noticeable, and the flesh melts away. I believe this is one of more sensuously flavored apple cultivars of late.

The core is smaller than most apples, which makes Ambrosia a lovely apple to core, slice and dry. The flesh does not brown quickly, which is a culinary bonus for those who want to use the apple in salads. It is not esteemed as a cooking apple, but it might be good as a baked apple. That cinnamon infused core of sugar is what draws us to the baked apple, but with Ambrosia perhaps you would not need to use as much.

Ah, my apple is now only a slim core attached to a relatively long, thin stem — the leftovers of this "scientific" evaluation.

Ambrosia is a midseason apple but with good keeping qualities. It is grown as far north as Zone 4. Time will tell how far this newcomer will range, but this is likely near its limit. An interesting historical note is that Ambrosia was an errant seedling found in the orchard of Wilfred and Sally Mennell in Cawston, British Columbia, in the 1980s. Apparently it was the pickers who first noticed the qualities of this upstart. The cultivar was patented but, luckily for propagators, the patent has run out in North America. It has been an up-and-coming cultivar since its entry into the marketplace.

One could do worse than to sink one's teeth into a vessel that holds ambrosia, the drink of the Greek gods, here stored in the cells of a humble apple — a mere mortal, whose lure, with the aid of a grafter, can keep it immortal. ■

Its eye appeal has helped Ambrosia rise to the top of cultivated apples.

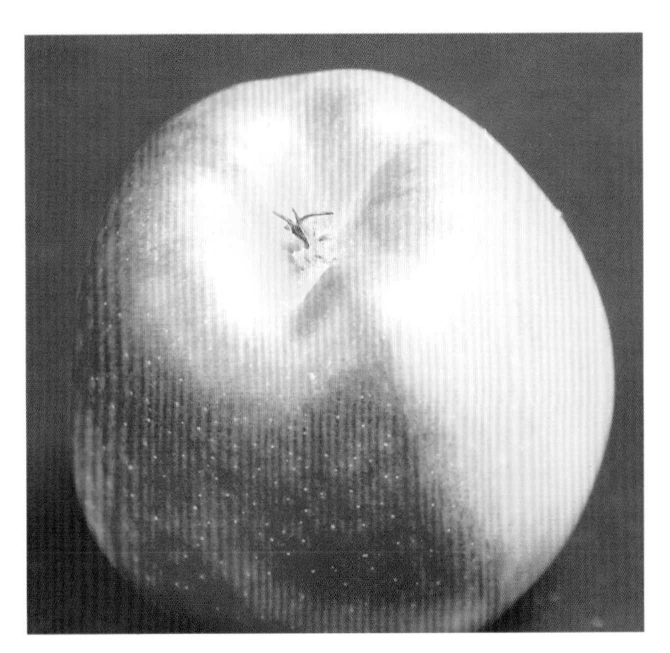

Ambrosia's basin shows five knobs.

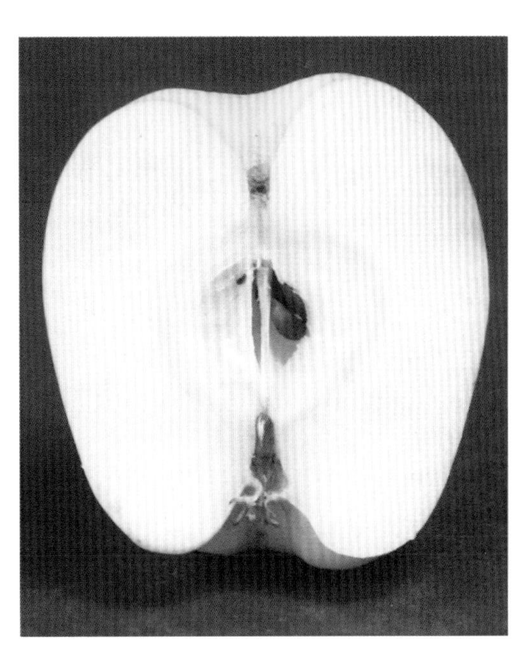

Ambrosia has a smaller core than most, with flesh that browns slowly.

Antonovka

Hardiness Zone 2
Introduced 19th century
Origin Kursk, Russia
Primary uses Cooking, baking, dessert

I have a special isolated orchard that only contains two cultivars, Beautiful Arcade and Antonovka. These are both old Russian apples whose seeds grow into seedlings that are highly valued by those looking for rootstocks that have superior hardiness, induce precocity into the scion, cause high productivity, are well anchored and usually produce trees that are approximately two-thirds the size of a typical seedling. The reason I have isolated these trees is to ensure that a great percentage of the seedlings will be crosses between the two cultivars and will be more likely to maintain the best features of these special trees.

In a trial of hardy cultivar seedlings conducted in the 1950s at the Kentville Research and Development Centre in Nova Scotia, Beautiful Arcade seedlings scored highest for all the valuable traits mentioned above. As well, the seedlings tend to have wide, dense, spreading root systems, great for shallow soils. Though a wonderful source of rootstocks and quite pleasant to eat, Beautiful Arcade never became popular, as it ripens very early and breaks down quickly. Antonovka is one of the best-known hardy rootstock sources, but it is also a good apple in its own right. Though hardly grown for eating in North America, it is one of the most popular apples in Russia, Poland and other Eastern European countries.

Beautiful Arcade is another apple of Russian origin that produces excellent seedling rootstocks, similar in characteristics to Antonovka.

Antonovka is believed to have originated in the city of Kursk, approximately 217 mi. (350 km) southeast of Moscow. The tree was brought to the United States in 1870. It is an upright to slightly spreading tree that is naturally smaller than most. It bears every year and is healthy, being little affected by apple scab and not particularly attractive to apple maggot. The fruit will hang on into winter if not picked. Most amazing is the hardiness of the tree. It will survive into Zone 2, making it among the hardiest of cultivated apples.

Antonovka's handsome fruit has a soft, dimpled surface. This, combined with its soft, rounded ribs, makes the fruit look somewhat like an overstuffed bean bag. It is round to oblate with a large, closed calyx set in a wavy basin. The stem is short to medium in length, nestled in a deep cavity that often has streaks of russet emanating from it. The fruit itself is solid green, which yellows with maturity. Inside is a flesh that is coarse, juicy and tart, though it has plenty of sugar for balance.

Antonovka can have a lumpy, sometimes asymmetrical form with characteristic russeting at the stem end.

Antonovka is a gift to the bakers of the world. Its acidity does not fade — allowing even stored specimens to produce wonderful baked products — yet its sugars are high enough to reduce the amount of added sugar needed for pies and baked apples. If picked when the skin's yellow hue has overtaken the green, it makes a lovely eating apple — refreshing and sweet at the same time.

If you are living where a walk in the winter without gloves is a shortcut to the emergency room's frostbite unit, this apple can provide a fine-quality fruit without any special attention. Paired with another super-hardy cultivar for pollination, it will enable you to participate in the delights of apple growing even while living in a land better known for short willows and groves of poplar. If you are a grafter, the seeds will produce deep rooted super-hardy rootstocks. It has also been instrumental in giving scab resistance to many of the newly bred disease-free cultivars. Antonovka is a very important apple, indeed. ■

Ashmead's Kernel

Hardiness Zone 4
Introduced 18th century
Origin Gloucestershire, England
Primary uses Dessert, cooking, baking, cider

This rather humble-looking rusty-brown apple was grown by Dr. William Ashmead of Gloucestershire, England, around 1700. It is most likely a seedling of the older cultivar Nonpareil. The doctor understood that a first-place winner for flavor can hide in a last placer for symmetry and shine. He certainly did the apple world a favor. Even after 300 years this odd little apple is considered one of the finest by those who enjoy a sprightly fruit with an array of complex flavors, the strongest being hints of pear. Its qualities also place it among the top cider apples.

Ashmead's Kernel has a background of greenish yellow but is often speckled or covered by a golden-brown skin. A reddish-bronze blush is common where the apple has been exposed to higher light levels. The form is round to oblate, but some specimens will be more conic. The apple is often softly ribbed and asymmetrical. It has a short slender stem and a rather shallow wavy basin.

Ashmead's Kernel is surprisingly tough for an English cultivar. When brought to the Americas it did exceptionally well in the apple growing belts, even in relatively cold areas. It can be grown in Zone 4, and some claim that on very hardy roots they are growing it in Zone 3 (though this may prove to be wishful thinking in the long term).

The tree is a bit dense and needs attentive annual pruning to help prevent it from bearing biennially. Branch and spur thinning will not only help provide more even fruiting, but also improve fruit size. Ashmead's Kernel on seedling roots may take five years to begin bearing, but on dwarfing stocks it will bear earlier. It is refreshingly healthy, showing good resistance to both apple scab and powdery mildew. The apples can tend toward bitter pit, so keep the soil pH at 6.5 to 7 and maintain good levels of calcium to prevent this symptom.

It is a rare roadside stand that will have baskets of Ashmead's Kernel for sale, and that is such a pity. We no longer have dozens of cultivars to choose from. How much more interesting would a trip to the market be if it were more like the bustling rural markets of England in the late 18th century, which would surely have had several grocers hawking Ashmead's Kernel to their discriminating customers? ■

Bailey Sweet

Hardiness Zone 4
Introduced ca. 1800
Origin Perry, New York
Primary uses Cooking, dessert

There was a humble third-generation beekeeper who lived in the village near my nursery in New Brunswick. When I had just started propagating apples I visited him, as his knowledge of hardy fruits was legendary in the area. I was not disappointed. He led me through his old family orchard, allowing me to gather scions of many historical cultivars that, by then, were in declining health. It was from that gathering that I started my Bailey Sweet. It is a relatively small tree even after 40 years, but it gives us a crop every year without fail. Spencer Ambrose Beach, in his book *The Apples of New York* (1905), wrote that Bailey Sweet did not excel in hardiness. Yet my tree, grown on Beautiful Arcade roots, has never suffered winter injury.

The Illustrated History of Apples in the United States and Canada (2016) by Daniel J. Bussey gives us an interesting history of this apple. It was first grown around the town of Perry in Wyoming County, New York, sometime around 1800, probably called Patterson Sweet but also known as Chillicothe Sweet and later as Bailey Sweet. It was grown in scattered locations from then on, but it was never an important commercial apple. Most were grown in home orchards by those who favored sweet apples.

The fruit is round to slightly conic and medium to large in size. The smooth skin starts as a clear yellow, becoming nearly completely blushed with deep red, and developing stripes and patches of darker red with noticeable white lenticels. The stem is short to medium in length, and the cavity is rather deep and somewhat furrowed. The basin is quite shallow and also slightly furrowed. The fruit and leaves are susceptible to scab, and precautions will need to be taken if you want clean fruit.

This is not an apple for most. It is perhaps the sweetest apple I have sampled, with little acidity for balance. Yet it is tasty in its own way, and its light-yellow flesh is juicy and moderately crisp. I believe Bailey Sweet is yet another unique taste that should at least be sampled by the jaded palates of today. A sweet blast from the past. ■

Ben Davis

Hardiness Zone 3
Introduced 19th century
Origin Unknown
Primary use Cooking

Truthfully, I did not want to include this apple. In the north it has been claimed that Ben Davis is only good for throwing at randy cats, and I have to agree. (Sorry, cat lovers, of which I am one.) Yet this apple was undeniably important in the orchards throughout North America in the 19th and early 20th centuries.

The very hardness that makes it a good projectile makes it a great keeping apple, and, more importantly for early American orchardists, it has a thick skin that made handling and shipping problem free. The origin of Ben Davis is hotly disputed. It makes its appearance in the early 1800s. Some believe it was found on the farm of one Ben Davis on the Nolichucky River in Tennessee; others say it originated in North Carolina; still others claim it is from Virginia. John Bunker of Maine has even speculated it may have originated in Maine, where it was commonly grown, and brought south. Whatever the truth, the fame of this apple spread across a good portion of the continent.

The tree is extremely hardy, even doing well in Zone 3, so the geographic reach this apple attained is extraordinary. In the south Ben Davis was touted as being well flavored, but in the north, where the fruit did not reach full maturity, it was regarded as rather tasteless. However, its firm flesh and skin still made it a hit with growers and, until the late 19th century, Ben Davis was one of the top cultivars in North America.

Ben Davis ripening in July.

The apple is medium to large in size and round to slightly conic. It can be somewhat elliptical. The skin is very tough and waxy with a glossy appearance. The background is greenish yellow, but it becomes bright red with darker vertical striping. Seedlings and sports of Ben Davis have even deeper color. Gano is perhaps the best known of these. My nursery has grown Gano for nearly 40 years. It has remained untouched by cold, even when the temperatures drop to −40°F (−40°C), and it is as hard and tasteless as Ben Davis. I always believed Gano to be a sport, but apparently the apple sleuths are fairly convinced it was a seedling of Ben Davis found by W.G. Gano in Parkville, Missouri.

There is no incentive to grow Ben Davis in the north unless you are a history buff. However, we cannot discount the impact this tough apple had on North American orchards of the past. ■

Ben Davis is a very hardy and healthy tree.

Gano is a seedling of Ben Davis that resembles it so closely it was thought to be a color sport until recently.

Bethel

Hardiness Zone 3
Introduced 1855
Origin Bethel, Vermont
Primary uses Cooking, baking, dessert

My first introduction to this apple took place on a spectacular steep hillside overlooking the broad reaches of the lower Saint John River. The site was an old orchard that hugged the rugged but fertile slopes of the mighty waterway. Many historical apples grew in this orchard, but the owner pointed to one as his favorite. "That's a Bethel. They don't get any better." I clambered into the giant old tree, worked my way to the top and clipped off several scions. I have grown Bethel ever since. While it is a matter of opinion whether it is the "best," it certainly is one I enjoy.

Bethel originated in Bethel, Vermont. It is most likely a seedling, or at least a relative, of Blue Pearmain, an apple it closely resembles. The first record of Bethel is in 1855. The fruit is round to slightly oblate, with some specimens a bit conic. It has a very short stem and a thick skin that is mottled and striped red, becoming darker with purplish-carmine tones and a bluish bloom at maturity. The color combined with its conspicuous russet and light-colored lenticels make identifying this old-timer easy. Another helpful hint when trying to identify this cultivar is to examine the buds on the new wood. They are deeply set into the stem, so much so they are barely noticeable.

The flesh is yellow with a hard, crisp texture and a satisfying quantity of juice. It is easy on the taste buds, neither too acidic, nor too bland. What sets this apple apart is a taste that is difficult to put into words. Suffice to say it is unique. This late-season apple will hang on the tree until season's end and will keep into winter. It is grown for fresh eating but is also esteemed as a good cooking and baking apple.

The tree is exceedingly healthy and hardy, and will grow to great size unless grafted onto a dwarfing rootstock. It is upright, though becoming quite broad and spreading with time. Although not immune to apple scab, it is relatively healthy and will produce a sizable crop annually. An interesting side note is that it has extremely long leaves.

There should always be a place for apples that are distinct. Bethel is as wonderful and tough as the people living in the state from which it hails and the people who have cultivated it since the early 19th century. Last I heard, the old hillside mother tree is still alive and well. ■

Black Oxford

Hardiness Zone 3
Introduced 1790
Origin Paris, Maine
Primary uses Dessert, cooking, cider

We can thank John Bunker of Maine for reviving this old apple. Its history dates back to 1787, when it was discovered in Paris, Oxford County, Maine. Once popular in the southern portions of the state, it was virtually extinct when Bunker encountered a bushel of Black Oxford brought in by a local farmer. He became driven from that point, as he notes in his book *Not Far From the Tree* (2007), to repopulate the earth with them. They are gracing the orchards again in Maine and starting to find adherents elsewhere.

Yet we cannot discount the comments from past growers in Maine. Its detractors have said it was "not a good cooking apple, was inclined to overbear, the fruit lacked character, was likely to be small and the leaves fell prematurely" (from *The Apples of Maine* [1993] by George A. Stilphen). On the other hand its adherents have said it is hardy and productive, good looking, a late keeper and free from apple maggot. I find that negative opinions of apples, when carefully examined, often reflect judgments made when the apples were not picked at the optimal time or perhaps were not allowed to properly ripen off the tree. The soil and geography where the apple is grown may also influence taste and keeping ability. Some of these may be the reasons for the conflicting opinions concerning Black Oxford.

In spring the blossoms are soft pink. The apple is round to oblate, sometimes a bit conic, with a long slender stem that emanates from a deep russeted cavity. The basin is shallow and wrinkled. The skin's background is yellow and becomes completely covered by a solid layer of dark purple-red with a bloom that gives the impression of near black. There are numerous white and russet lenticels. It reportedly has some resistance to disease and insects.

The dense, hard flesh transitions from somewhat vegetative to rich and sweet, but with enough acidity to be in balance and with subtle vanilla and tropical fruit notes. This is a very late apple, so it should not be picked until in danger of hard frost, which is in early November at my site. It will keep extraordinarily well until spring, becoming mellower with time. Sources in Maine say it makes an excellent late cider.

Black Oxford is among the darkest of apples, though the cultivar called Tangowine is close. Another new apple called Black Pearl was released by the Kentville Research and Development Centre in Nova Scotia. It is also exceedingly dark.

We owe a great debt to the rather esoteric group of people who feel driven to save trees like Black Oxford from oblivion. It seems that when there is a need to rediscover disappearing skills or, in this case, foods, people appear seemingly from nowhere to keep these important things alive. ■

Blue Pearmain

Hardiness Zone 4
Introduced Late 18th century
Origin Middlesex County, Massachusetts
Primary use Dessert

I include this old-timer out of respect for its historical importance. Blue Pearmain was believed to have originated in Middlesex County, Massachusetts, in the late 1700s. It spread from there to become an important apple in New England, being planted as far north as Ontario and the provinces of Atlantic Canada. Its main claim to fame was superb hardiness and an ability to keep in common storage well into the winter.

Blue Pearmain is an unusual-looking fruit. It is medium to large and round to oblate, sometimes conic. It has a short, thick stem. Its surface can be slightly ribbed or even furrowed. The skin is a random collage of red and yellow with patches of russet near the stem cavity and scattered across the surface of the fruit. As it matures, the apple appears more purplish red overlain by a bluish bloom, giving it an effect somewhat like that of a big striped grape. Its lenticels can be either small or large gray dots with a russet center. The closest in looks to this apple is Bethel, which probably owes its genes to this earlier cultivar.

Blue Pearmain is a hard, crisp apple when picked, which usually occurs in October in northern areas. Straight off the tree it will be tart, gradually mellowing as it ages. In his marvelous book *Apples of Uncommon Character* (2014), Rowan Jacobsen describes Blue Pearmain's aroma as "a big burst of musky orange, flanked by a hard-charging kiwi brightness and a tropical melon finish."

Though I am in awe of the descriptive skills of this pommelier, many have found this apple wanting in the quality department and, more importantly, have found it rather unproductive. As the 19th century progressed, fewer Blue Pearmain trees were planted, and now only a few of the older trees remain. There is, however, a new generation of propagators that is keeping this cultivar from falling off the edge into oblivion. Certainly for the collector, a single tree would be in order. ■

Bonkers (NY 73334-35)

Hardiness Zone 4
Introduced Early 1990s
Origin Geneva, New York
Primary use Dessert

Sometime around 1995 my nursery ordered several numbered cultivars from the New York State Fruit Testing Cooperative Association. These were selections from the breeding program at Cornell University in Geneva, New York. We were especially interested in their disease resistance. Although we observed them for several years, it seemed that none were to be introduced, so we abandoned hopes of putting them in our propagation lineup.

Suddenly I heard about this apple with a name that is bonkers, literally. Michael Phillips, who runs Lost Nation Orchard and is the author of several books on organic apple growing, had fallen in love with one of the selections, NY 73334-35, and had unilaterally named it Bonkers.

The apple is a cross between Liberty and Red Delicious. The tree is nearly impervious to apple scab, has good vigor and is also parthenocarpic, meaning it does not have to be pollinated in order to set fruit. This is similar to fruits like seedless watermelons and grapes. The result is a large, often seedless fruit with a near-nonexistent core. Bonkers, eh?

The apple itself is oblate and often asymmetric, which has a lot to do with its parthenocarpic nature. The stem is medium in length. The skin has a background of soft yellow. As the fruit matures, a few patches of the background color remain, but most of the surface blushes a deep red, with some areas turning a deep purple-red. There are sparse but prominent white lenticels scattered over the surface.

The flavor of Bonkers has been described as similar to Liberty but with a bit more tartness and hints of pineapple. The flesh is creamy white with a good crunch and plenty of juice.

After reading about Michael Phillips's glowing endorsement of this apple, I decided to wander back to the old stock block where it was planted long ago. These trees are cut back quite hard every two years or so to produce scion wood, though none had been taken from the New York selections for many years. Sure enough, we still had a tree of NY 73334-35. It is actually the largest and most vigorous of all the selections. You can bet we'll be taking scions from it this year.

In truth, I had always been partial to another New York selection, NY 75413-30. It is a huge apple that ripens in early November. Maybe I should get into the name game. ■

Bottle Greening

Hardiness Zone 3
Introduced 1866
Origin Vermont
Primary uses Cooking, dessert

I slip this old and now little-known apple into the mix because it is a hardy, healthy great cooking apple with an "intoxicating" story behind it.

Bottle Greening was a seedling found growing in a Vermont orchard near the New York State line in the early 19th century and first recorded in 1866. Through the years the pickers used the tree, which had a convenient hollow in the trunk, to hide their bottles, perhaps from a teetotaling owner. But this tree provided the apple world with much more than a rustic liquor cabinet. A man named Eben Wight of Dedham, Massachusetts, was impressed by the tree's virtues, and soon its wood was being passed from grower to grower across a portion of North America.

Bottle Greening possesses equal or better hardiness than its probable parent Rhode Island Greening. It is upright growing, becoming more pendulous with age, and is said to prefer open soils. It has certainly done well on our shale-based soil. The tree flowers fairly early in the season and is a reliable annual producer of oblate to slightly conic grass-green fruits, sometimes with a soft-yellow cast and a dull pinkish blush when exposed to full sun. The lenticels are barely noticeable, which gives the fruit an even coloration and "clean" look.

The flesh is white and firm but tender and juicy. It is mildly aromatic and has good acidity, which makes it a very fine cooker, as well as a good eating apple. It also has a small core. The skin is easily bruised, so the fruit should be picked with care. This is one of the reasons given for why Bottle Greening lost out to Rhode Island Greening in the "battle" for green cooking apple prominence.

The oblate fruit is grass-green ripening to dull green at maturity.

I include this apple to heighten awareness of a nearly extinct apple that is an excellent addition to the northern orchard. A few nurseries still list it, but it is becoming hard to find. Perhaps an entry like this will help keep it from becoming extinct. ■

Most likely a seedling of Rhode Island Greening, Bottle Greening has all the best attributes of that old cooking apple, plus better hardiness.

The tree ends up quite rounded. Beneath it is a year's prunings.

Bramley

Hardiness Zone 4
Introduced ca. 1846
Origin Southwell, England
Primary uses Cooking, baking

Few apple histories are as well documented as that of Bramley (alias Bramley's Seedling). This famous cultivar originated as the whim of a small girl, Mary Anne Brailsford of Southwell, Nottinghamshire, England. She planted several pips (seeds) in pots. One grew well and was put in the family garden in 1809. It grew into a vigorous tree with large green apples. The family moved and the cottage was purchased by a local butcher (though one source says shoemaker) named Matthew Bramley in 1846. In 1856 the nurseryman Henry Merryweather asked to propagate and sell the tree. Bramley agreed, with the proviso that the apple be named after him. By 1876 it had been highly recommended by the Royal Horticultural Society and up until recently accounted for nearly all the cooking apples sold in England. Although blown over by a storm in 1900, the original tree still survives as of writing this book, though it is dying of honey fungus. The tree is considered a national treasure. The pity of it all is that this apple was not named Mary's Pippin.

This apple's notoriety is due to its excellent culinary characteristics and extremely large size. It is a somewhat blocky round fruit, often ribbed and grass-green with red striping where it receives sun. It has a short thick stem. If picked before fully ripe, Bramley is quite sharp, but mellows somewhat as it ripens, which occurs around mid to late October in New

Bramley's a clean green apple until maturity.

Brunswick. The flesh is white, often tinged with green. The fruit has the unusual quality of becoming golden and fluffy when cooked. This characteristic, combined with its acidity, makes it one of the finest apples for pies and pastries. It will keep in cold storage till midwinter.

The tree is stout and wide, reflective of its twigs and branches which are much bulkier than most. Indeed, this is one of the most pleasing tree forms for me, though picking is a challenge as the branches tend to be low and horizontal. Bramley is quite strong and has done well in Zone 4. Both fruit and leaves are very resistant to apple scab. This is a triploid apple and requires the pollen of two other cultivars for pollination.

In an age when nearly all apples sold are dessert apples, it is often difficult to find fine cooking apples. If you have a chance to purchase this fruit do so; if you have a chance to plant a tree, even better. ■

The tree is wider than nearly any other cultivar, with thick branches.

The most popular cooking apple in England, Bramley is known by few in North America — a pity.

Carroll

Hardiness Zone 3
Introduced 1961
Origin Morden, Manitoba
Primary uses Dessert, cooking

Although this apple will never be a North American wonder child, Carroll is the kind of apple that plays an important role where temperatures regularly drop below –40°F (–40°C). The apple was the result of breeding work done at the Morden Research and Development Centre in Manitoba. Carroll was released in 1961, a cross of 5029-EL52 (a seedling of Moscow Pear) and Melba.

The skin is a cream-yellow blushed with a solid red with little or no striping. It is an early apple that can be kept up to two months. The white flesh is crisp with a good tart-sweet balance that has more complexity and aroma than most super-hardy cultivars. Carroll is used for both fresh eating and for making pies and sauces.

The tree is vigorous and slightly higher than wide. It is reported to be resistant to fireblight and will survive a bone cracking –50°F (–46°C). ■

Chestnut Crab

Hardiness Zone 2
Introduced 1949
Origin Minneapolis, Minnesota
Primary uses Dessert, cider

Chestnut Crab is a large crabapple developed by the world-famous University of Minnesota breeding program as a seedling of Malinda and Siberian Crab, planted in 1909, selected in 1921 and released in 1949. Although it has taken a while to rise in stature, consider this a cult star today.

It is a round to oblate fruit with a long stem. The skin of a Chestnut Crab is yellow but becomes nearly all reddish bronze, often with a light covering of russet. The flesh is what keeps people begging for more and cleaning out the supply of any market that carries it. The texture is fine grained and crisp with some tartness, but it does not have the wild acidity found in most crabapples. The flavor is complex and has, what many call, nuttiness. Rowan Jacobsen in his book *Apples of Uncommon Character* (2014) describes it as "a peach pie on a graham cracker crust."

The fruit hangs well on the tree and is usually picked in early to midseason, early September here in New Brunswick. It will keep for up to one month, far longer than most crabs. It makes a tasty treat for child and adult alike, and it has been getting rave reviews from cider makers.

Lastly, it is a good pollinator in the orchard. It blooms midseason, so its flowers overlap with most cultivars. You will see some descriptions that say it is self-pollinating, but it is not. It needs the pollen of another cultivar. ■

To my taste, Chestnut is the tastiest of crabs.

When fully ripe, the fruit is covered in a web of light russeting. Note the long, thin stem.

Cortland

Hardiness Zone 3
Introduced 1915
Origin Geneva, New York
Primary uses Dessert, cooking

A freshly picked Cortland takes second place to few apples for flavor. It originated as a cross between Ben Davis and McIntosh in the trial orchards at the New York State Agricultural Experiment Station in 1898 and was released in 1915. In the northeastern United States, Ontario and Atlantic Canada this apple, until very recently, was one of the most heavily planted, usually as a pollinating companion for McIntosh.

Cortland has a distinctive flavor with perhaps more vinous qualities than its famous parent. The large, squat fruit is round, sometimes a bit irregular and with a red wash and striping over a yellow-green background. Newer color sports such as Redcort are a deeper red, more in tune with market demands. The flesh of Cortland is juicy but not overly crisp. It is pure white and has the remarkable attribute of not browning when cut, as most apples do. This has been part of the allure of Cortland for cooking and for salads, dishes in which its ability to stay white is greatly appreciated. It will keep until December in cold storage and till April in controlled-atmosphere storage.

Cortland is a large, rounded and spreading tree with a distinctive pendulous habit. The fruit is borne on the tips as well as on spurs. It is quite hardy and has been grown in many of the colder growing regions with good success.

Until recently Cortland was the number-two apple after McIntosh in the Atlantic provinces and much of New England.

It is ripe in New Brunswick in early October and can usually be harvested in one picking. Although it will hang on the tree until later, the fruit will become softer and the skin waxier. They will not store as well as those picked at optimal harvest time.

The list of diseases that attack Cortland is long. It is susceptible to apple scab, mildew, fireblight and cedar apple rust. It is resistant to canker, which in large part may be due to its hardiness and lack of winter injury. It is also resistant to collar rot.

Despite some faults, no list of excellent apples would be complete without this multipurpose cultivar. It is still a favorite with those who have eaten it all their lives. ■

The tree is quite pendulous as it ages. Fruit is often borne on the tips of the branches.

Unfortunately Cortland is quite subject to diseases such as scab,
but that should not deter you from growing this great cultivar.

Cosmic Crisp

Hardiness Zone 3
Introduced 2019
Origin Wenatchee, Washington
Primary use Dessert

I have never even tasted this apple, yet to leave it out of the book would probably be a mistake. This is an apple that took decades to produce. The original breeder, Bruce Barritt, who worked for the Washington State University breeding program, wanted to create the "perfect" apple — one with a crisp juicy texture, good flavor, good disease resistance, good production and good looks. Cosmic Crisp is the result of sifting through 10,000 seedlings that resulted from a cross between Honeycrisp and Enterprise.

The name apparently came from a tasting session. One of the participants said the lenticels resembled stars. Indeed, this round to conic apple is a solid red when ripe with nearly perfectly spaced white lenticels. A touch of yellow can be found in the stem cavity, but with enough sun, the entire fruit becomes a deep and shiny red. While names are important in today's marketplace, in the end it is the flavor and texture of the apple that will be its legacy. It would seem from initial reports that this is a truly exceptional dessert apple. Its crunch has been touted as the crunchiest of the crunchy. It also browns slowly, a boon to those who want to use it in salads, and the balance of sweet and tart is exceptional.

The hardiness of this apple has yet to be determined, but both its parents are hardy apples and so it is likely to be fine in Zone 4 and possibly Zone 3. So far I have little data about the tree and its characteristics. In other words, I am taking a leap of faith believing the intense hype this apple has received. Whatever the final assessment, there is little doubt it will soon be a contender for the coveted space in the apple section of the store.

This apple can only be grown by Washington State growers until 2027, so trees will not be available until after that date. Those of us outside of Washington will have to be satisfied with fruit in the stores till then. ■

Cox's Orange Pippin

Hardiness Zone 4
Introduced 1850
Origin Colnbrook Lawn, England
Primary uses Dessert, cooking

Praises for Cox's Orange Pippin have been sung ever since it was first introduced in 1850. It has, until recently, been the dessert apple of choice in England, where it is widely considered the pinnacle of flavor and the apple against which all others are judged. It was grown from seed of Ribston Pippin, which was most likely a seedling of French provenance that was planted with other seeds at Ribston Hall in Yorkshire, England. Cox's Orange was raised by a Mr. Richard Cox around 1800 in Colnbrook Lawn near Slough, Buckinghamshire, in England. The original tree blew down in a storm in 1911.

Cox's Orange is not the modern ideal of an apple. Its green background is shaded with orange to soft brick red and is often mottled with russet. It can have a warty appearance and is sometimes asymmetrical. Whatever it lacks in refinement on the outside, it makes up for on the inside. The flesh is fine grained, juicy, sprightly subacid and richly aromatic. Its complex flavor has been described much as a sommelier might describe a wine, with hints of nuts, spice, pear, mango, melon, peach and orange. In a word the flavor is fantastic, and for many adoring fans, unmatched.

What has been described, however, is not always what one purchases at the store. This apple is fussy in its needs. It prefers a maritime climate yet not one that is overly damp. It likes cool temperatures and abhors high heat. These requirements for perfection mean that many of the great apple-growing regions of the world cannot grow a good Cox's Orange. They are often bland and bear little resemblance to those grown in the drier areas of England, which have always been considered the ideal sites for this apple.

The tree is not easy to grow even under good conditions. Though it is considered relatively productive, Cox's Orange can be an irregular producer. It is only moderately vigorous and produces an upright dense canopy with slender crooked branching. It benefits greatly from careful detailed pruning. It is prone to canker; the cold temperatures in northern areas can cause winter injury, which provides entry points for the disease. Young trees will often produce fruit that cracks, though this usually lessens with age.

It is essential for the northern grower who wishes to test their skills to purchase trees grown on very hardy rootstocks. On such roots Cox's Orange can be grown in Zone 4 with some success.

Cox's Orange is not an apple for the timid grower, but if you have reasonably good conditions, it is certainly worth trying. You may wait a while to produce your first good crop, but if all goes well, you will be in for a treat. ■

Crimson Crisp

Hardiness Zone 4
Introduced 2006
Origin Cream Ridge, New Jersey
Primary uses Dessert, cooking

If you are a fan of a hard apple that snaps when bitten, you will enjoy the experience of eating this apple. If you are a fan of deep red-skinned apples you will be ecstatic with the look of Crimson Crisp.

This apple is one of many released from the PRI breeding program that involved Purdue University, Rutgers University and the University of Illinois. Starting with a selection of the *Malus floribunda* crabapple (clone 821), which has immunity to apple scab, the breeders crossed it with many well-known apples in the hopes of producing an apple with great-quality fruit and disease resistance.

Crimson Crisp, originally known as Co-Op 39, was the result of a series of crosses that culminated in a selection made in 1979 by E.B. Williams and finally released in 2006. I wonder how he reacted when he first bit into this dark red apple, one of so many he would have had to judge. Obviously he knew it had great potential.

Crimson Crisp is a medium-sized apple that ripens late in the season and can be kept in cold storage for up to four months. The fruit is ovate-round to roundish and very symmetrical. The stem is quite short. Other than a yellowish basin, the entire skin is a glowing deep purple-red. The basin tends to be open, and occasionally mold will invade the core.

The flesh is yellow and mildly acidic with a complex flavor, though not overly aromatic. It is snapping crisp with a lot of juice, perhaps on a par with the explosive juiciness of Honeycrisp.

The tree is small to moderate in size with nicely angled branches, though it tends to produce a profusion of branches, which may require some thinning to open up the tree. It tends to bear fruit singly, which means thinning is not as crucial. Another advantage of this apple is that it hangs on the tree, allowing it to be picked for nearly four weeks.

With many competitors for market share, this newcomer will have a tough time making it to the penthouse, but for those interested in disease resistance combined with a great crisp and flavorful mouthful, Crimson Crisp is a wonderful choice. ■

Delicious (Red Delicious)

Hardiness Zone 4
Introduced 1895
Origin Peru, Iowa
Primary use Dessert

Few apples have had the exposure Delicious has had. It is known throughout the world and has its adherents and detractors. It has become the classic apple form for many illustrators and photographers, and is ubiquitous on the supermarket shelves. However, there are many who bristle at the name and begin ranting about an apple that is unjustly overrated. My task here is not to praise, nor criticize but to lay out the facts.

It is a fact that this apple had a rough start on the farm of Jesse Hiatt of Peru, Madison County, Iowa. Delicious began life as an errant seedling in an orchard of Yellow Bellflowers. Twice it was cut down, and the third time it grew back Hiatt was said to have muttered, "If thee must grow, thee may." The persistent seedling began producing apples in about 1881. Hiatt was smitten by the form, color and taste of his maverick and named it Hawkeye. He was so excited that in 1893 he sent specimens of the apple to Stark Bro's Nurseries & Orchards Co. in Louisiana, Missouri. As the story goes the president of Stark Bro's, Clarence Stark, cried out, "This apple is delicious," and then said, "That shall be its name."

The Starks were quick to find Mr. Hiatt and buy the rights to the apple. Two years later it was introduced as Stark Delicious. In 1914 the name was registered with the U.S. Patent Office as Red Delicious, to be marketed in tandem with their other great find Golden Delicious. Stark Bro's believed they had a winner on their hands. They created tens of thousands of Red Delicious trees and shipped them to growers across the country. They also spent $750,000 over the next two decades promoting their new wonder child.

In 1923 a New Jersey grower reported he had found a single limb on a Delicious tree that had changed color earlier and was far darker than the rest of the tree. Paul Stark, Clarence's son, visited the grower and paid him a tidy $6,000 for the limb. From then on sports such as this helped change the look of the original red-striped Hawkeye, morphing it into a flashy red fruit that shone like a crimson jewel in the produce sections of American supermarkets. It is interesting to note that by favoring a redder, thicker skin over the original yellow, striped skin of Hiatt's Hawkeye, growers and propagators effectively diminished Delicious's original flavor and overall palatability. This is because the genes for flavor-producing compounds are found on the same genes as those for yellow skin.

By the mid-20th century Delicious was the number-one apple produced in the U.S. Washington State quickly became the country's biggest apple-growing region because growers there could create a near-perfect Delicious. In fact, Washington State growers had so much faith in the cultivar that by the 1980s 75 percent of crop production in the state consisted of Delicious. Until recently it was the best-selling apple in America, and it found its way around the world due to increased exports from the U.S.

Here are Delicious grown in northern conditions. The fruit is usually a bit smaller with denser flesh and lower sugar.

Delicious is a medium to large fruit with a classic oblong-conic form. The glossy pale-yellow background of the young fruit becomes overlain with dark red striping and splashes of crimson red. The skin is somewhat thick, and there are no patches of russet to spoil the glossy sheen. The ivory-white lenticels are scattered evenly but are not prominent. The basin is knobbed, and the opposite end hosts a stem that is medium in length, fairly stout and curved. The flesh that causes heated arguments among so many is yellow, crisp, with a light acidity that is overshadowed by sweetness. Delicious is good quality for dessert but is generally not used in cooking.

One of the negative reactions against this fruit is that it is common, and some people simply cannot abide common. However, part of this criticism may be explained by improper picking times. Many is the time that you bite into an inviting-looking Delicious and inside the flesh is nearly green — enough to turn anyone against the apple. One reason Delicious is a favorite among growers is for its shipping

Here are classic Delicious grown in Washington state. They mature better in a warmer climate.

ability, so often the crop is picked for shipping when not completely ripe. If picked off the tree at the proper time, which is early November in Zone 4, it is more likely to at least try to live up to its name.

The tree is only moderately vigorous with slightly zigzagging branches that give it a wayward appearance. It is relatively precocious and produces a good crop every year. This is another reason it remains a favorite of growers. Although not the worst for scab, it is susceptible, so prevention is necessary to produce quality fruit.

Another factor for those living in cold areas is the fact that Delicious needs a good deal of heat to properly ripen. For instance, in our orchard we end up with smaller fruits that have good color but a dense flesh that cannot compete with Delicious grown in warmer climes.

Honestly, this is not an apple I would recommend to the northern grower, but you simply cannot disregard Delicious and the immense impact it has had on the world of apples, even in colder areas. This giant among apples, however, has seen its share of the market drastically reduced of late. Newer cultivars such as Gala, Honeycrisp and Ambrosia are chipping away at its prominence, and it will undoubtedly play a much smaller part in the future apple mix of North America. ■

Dolgo

Hardiness Zone 2
Introduced 1917 (in the United States)
Origin Russia
Primary use Cooking

There was once a huge demand for crabapples. Before widespread refrigeration and modern storage facilities, canning was a popular method of preserving apples. Canning whole crabapples and making crabapple jelly was a yearly rite among the rural population.

Over the past century the crabapple has been gradually relegated to the back shelves of apple production. However, some enterprising growers still keep a few trees for the folks who continue to put down several bottles of pickled crabs a year or who cannot live without crabapple jelly. It is encouraging to see a new generation of preservation artists coming onto the scene. As well, there is new demand for the cider crabs that, until recently, were sliding toward extinction.

Classic Dolgo at complete ripeness. Note the oblong shape and long, thin stems.

There are innumerable crab cultivars, many of which are grown in the far north because of their superior hardiness. Rescue, Trailman, Shafer, Siberian Red and others are still available from a few specialty nurseries. The once widely grown Transcendent crab has become another casualty of modern horticulture. This hardy crab has an unmistakable form. The trunk, often gridded by the holes of sapsuckers, becomes huge and towering, often to remarkable heights, yet with branches that bend downward in graceful arches — a sort of weeping willow effect. The ovoid fruit, a rather dull yellow blushed red, was once the crab of choice for many a farmstead. However, Transcendent, like many of its fellow crabs, suffers from the scourge of apple scab. In a bad year scab can make much of the fruit unusable, or at least unappealing for pickling.

In 1917 Niels Hansen, a breeder working at the Agricultural Experiment Station in Brookings, South Dakota, introduced a new crab he had obtained from Russia in 1897, probably a seedling of *Malus baccata*. He called it Dolgo.

Dolgo is extremely hardy, doing well in Zone 3 and probably Zone 2. Not only is it tough, but it is also very healthy, being resistant to both apple scab and fireblight. The tree is vigorous and upright, nearly vertical at first but gradually forming a wide ovoid form, reflective of its fruit. It has distinctive new branches that

Dolgo ripening on the tree in July. It is a good idea to pick Dolgo a bit on the unripe side if you are going to make jelly.

are greenish yellow with reddish overtones. Not only is Dolgo a good eating and preserving cultivar, but it is also used in many orchards to provide pollen, which it produces in profusion.

The fruit is flavorful and of good size for a crabapple. Its tartness is toned down by a sweetness rarely found in crabs and a flavor that has hints of pineapple and strawberry. Though most of its fruits find their way into the jelly pot, many people, myself included, enjoy the fruit right off the tree. The oblong crabs are initially yellow but gradually blush red till they are fully red at maturity. Picking before they complete their color change is ideal for preserving, as they will be tart enough for great jelly. Once fully red, they will soon become mealy.

Dolgo has proven to be a survivor among crabs. Its many attributes have saved it from the near oblivion of its kin. It still holds a place in the orchards of today and we can only hope it will continue as a fine ambassador of the small-fruited apples. ■

Duchess (Duchess of Oldenburg)

Hardiness Zone 2
Introduced 19th century
Origin Russia
Primary uses Cooking, dessert

Duchess of Oldenburg, now usually called Duchess, was brought to England from Russia around 1825. Ten years later it was one of four Russian apples imported from England by the Massachusetts Horticultural Society, the others being Alexander, Red Astrachan and Tetovsky. Duchess quickly found favor, and its cultivation spread northward through New England to Atlantic Canada and westward to the Mississippi River, where many earlier introductions had succumbed to extreme winter temperatures.

This is an early apple, and if picked at peak, it provides a juicy treat with high marks for aroma. It has a short shelf life and has been used primarily as a culinary apple. The white to yellow flesh is slightly tart, making it ideal for pies and sauce.

One of its most endearing attributes is resistance to apple scab. Though not immune, most years the incidence of infection is low for an early apple. The tree is bone hardy and will easily survive temperatures of –40°F (–40°C). The shape of the tree is upright but widening with age. It is my experience that the trunks remain somewhat more slender than those of similarly aged cultivars. Crops are biennial, though some cropping occurs even in off years.

This is one of the gems of our hardy old apples. Where I live it was once planted in every orchard. One is enough in most orchards as the fruit does not keep, but the flavor will make you eagerly anticipate each year's crop. ■

Duchess nearly ripe in late July. This is a very early apple that is excellent for cooking.

The tree is smaller than most and upright to arching.

Dudley (Dudley Winter)

Hardiness Zone 3
Introduced 1888
Origin Castle Hill, Maine
Primary uses Cooking, baking, dessert

You cannot get too much farther north in the contiguous United States than the small town of Castle Hill, Maine. Despite the name, Castle Hill is quite flat with good soils that grow mostly potatoes. It was here around 1875 that J.W. Dudley grew seedlings of Duchess, one of the hardiest apples available at the time. The one he chose, supposedly a cross with Hyslop Crab, came to the attention of Maine growers in 1888. Specimens were shipped to the Chase Brothers Nursery of Rochester, New York, and it was introduced as North Star. However, it was discovered that there was already an apple named that, so the name was changed to Dudley Winter.

This apple spread rapidly through Maine and across the border into New Brunswick and was being grown in places such as Wisconsin. Within a short time Dudley had become one of the most important apples in the north.

Perhaps the strongest proponent of Dudley was Francis Peabody Sharp of Upper Woodstock, New Brunswick. He owned the largest nursery and orchard in Eastern Canada at the time and had been so impressed with Dudley he began producing thousands of trees, noting its improvements over existing cultivars.

The tree was as hardy as Duchess and sturdier than its parent, being less liable to splitting. It produced fruit early in life and was annual. Perhaps most importantly, Dudley was a late apple in the north, ripening in late September/early October. It kept well till late winter with little loss in quality.

The medium to large fruit is yellow-green that becomes splashed and striped with pastel pink to soft red. It develops a bluish bloom. It is round to oblate with a long, thick stem in a cavity that can exhibit streaks of russet. It is briskly acidic but with enough sugars to make it an enjoyable snack. Its real value, however, is as a baking and sauce apple.

Eventually the introduction of newer cultivars reduced the numbers of trees grown. There are still orchards that grow this old-timer, but fewer each year. Dudley is an excellent choice for cold climate growers, but it is rarely sold today. Nevertheless, it remains a valuable apple that is loved by those who know it. ■

Empire

Hardiness Zone 4
Introduced 1966
Origin Geneva, New York
Primary uses Dessert, cooking

Empire is one of many apples that were bred using McIntosh as the mother. In this case the pollen parent was Golden Delicious. The original cross was made in 1945. The first fruit was produced in 1954 and was eventually named in 1966 in honor of the Empire State (New York). It takes a long time to create a new cultivar.

Although it took many years to bring to fruition, the breeders at the New York State Agricultural Experiment Station knew they had a winner. The fruit is medium in size, round to oblate-conic and is often irregularly shaped. The skin develops a deep, glossy purplish red over nearly the entire surface when sun drenched. White lenticels are prominent and scattered evenly across the fruit. The stem is long and slender. Although the outside is attractive, the best part lies inside — a flesh that is creamy white, crisp and juicy with excellent quality. Although not impervious to scab it is relatively clean and resistant to cedar apple rust.

The tree is a grower's idea of perfection, forming well-spaced, strong branches that are easy to prune. The fruit is evenly distributed throughout the tree. Empire is a dependable annual producer, and it produces good crops, more so than Red Delicious, the number-one commercial apple of the 20th century. It ripens two weeks later than McIntosh and keeps well.

Empire remains most popular in its home state, though it is grown in many orchards throughout New England and occasionally farther afield. In the north the problem is the length of season required to properly ripen it. For example, in New Brunswick it will ripen if we have enough warmth in the year. So it is worth growing in Zone 4, but I would hesitate to recommend it for colder zones. ■

Enterprise

Hardiness Zone 4
Introduced 1993
Origin West Lafayette, Indiana
Primary use Dessert

Other than hardiness, disease resistance has been my priority when seeking out new cultivars. I have always taken an interest in apples introduced through the Purdue University, Rutgers University and the University of Illinois (PRI) breeding program. Enterprise is a relatively new introduction, being released in 1993 as a cross between PRI 1661-2 and PRI 1661-1.

The tree is spreading, round topped and moderately vigorous. It bears its apples singly and evenly throughout the tree. The fruit is medium to large, round to oblate, often becoming more elongate as the tree matures. Enterprise becomes nearly completely covered in bright red with hints of yellow and orange where it is shaded. Every indication is that it should be hardy in Zone 4 and possibly Zone 3; however, because it is quite late (two weeks later than Delicious), colder areas probably will not be able to bring it to optimal quality. Complete maturity might even be a problem in Zone 4.

The flesh is pale yellow, firm, crisp and breaking with rich, spicy overtones. Enterprise has a high level of acidity at harvest but mellows in storage. It is rated as an excellent-quality fruit that stays firm and crisp even after six months in storage.

Enterprise is a high-quality dessert apple that is starting to get widespread exposure, and it is being used as a parent in new breeding work. It could be of importance to organic producers, as it is immune to scab and very resistant to cedar apple rust and fireblight as well as moderately resistant to powdery mildew. Bitter pit and watercore have not been problems for this cultivar. ■

Fameuse (Snow Apple)

Hardiness Zone 3
Introduced 17th century or earlier
Origin Unknown
Primary uses Dessert, cooking

There are many theories as to the origin of Fameuse. Most believe it was grown from seed brought to Quebec (then known as New France) by missionaries around 1600. There is an old apple in France called Pomme de Neige (Snow apple), and perhaps this was the original source of Fameuse's seed. It may also have been brought as a grafted tree from France by an obsessive horticulturist. Though less likely, it is certainly possible.

There is a newer theory for its origin. After visiting the Basque Country in Spain and seeing a similar apple growing there, orchardist and apple historian John Bunker has postulated that it may have been brought to Newfoundland as seed in the 1500s by Basque whalers. The Basques grew apples for cider, with which they filled the holds of their ships on their way to the shores of Newfoundland. They could have brought seeds from the orchards of home and grown small orchards around the whaling stations to produce cider and keep the whalers happy. Fameuse produces seedlings that are very similar to itself, and their quality might well have attracted the attention of Jesuit missionaries stopping on their way to New France (Quebec). Perhaps they collected seeds from these apple trees and brought them to Quebec, where they were planted in the missions and later in the newly cleared lands along the St. Lawrence River.

Fameuse in August. They are an icon of the oblate form. Note the dimple caused by a tarnished plant bug.

When fully ripe Fameuse is one of most satisfying flavors of any, and the color is burgundy red.

Fameuse spread rapidly west and south from there. By 1700 it was being planted at Chimney Point on Lake Champlain between New York and Vermont. In fact many attributed its source to Chimney Point, Vermont, where it was called the Chimney apple, but it had been grown throughout

Fameuse trees can grow very large on seedling roots.

most of the 17th century in Quebec before it arrived in Vermont. Nevertheless, it soon became one of the most popular apples in the north.

Fameuse is a small- to medium-sized apple, although larger specimens occur. It has a classic oblate form, sometimes becoming roundish. The apple has a tender skin that may be blushed and striped red when grown in shade, but it develops a deep bright red coloring in sun. The stem is usually short, and the apple has a deep stem cavity that is gently furrowed and whose surface is often textured with russet. The basin is small, often with protuberances.

Fameuse's flesh is pure white, often with delicate streaks of red that emanate from the skin. The fruit is tender, juicy and sprightly but sweetens as it matures. It is quite aromatic and makes a great dessert fruit but can be used for cooking as well.

I have always thought, as have many others, that Fameuse is the parent of McIntosh. I have seen striking similarities between it and Novamac, a child of McIntosh (and a would-be grandchild of Fameuse). However, I recently received information that DNA testing has debunked this theory. I still await confirmation.

Fameuse was one of the most popular cultivars of the 18th and 19th centuries in the northeast and was still being planted into the early 20th century. However, as agriculture industrialized, there was no longer room for smaller fruit. Large fruit produces better packout, which translates to better profit — fewer fruits to fill the bins. Most Fameuse trees exist only as last holdouts in old orchards, and they can be massive. Mine are towering giants, among the largest trees in the orchard. Thankfully there are those who will not accept the demise of this wonderful and historical apple. The grafters of today are ensuring there will always be some to tempt the jaded palates of the future. ■

Fireside

Hardiness Zone 3
Introduced 1943
Origin Minnesota
Primary uses Dessert, cooking

Here is a Minnesota-bred apple that has never had its due. While popular in its home state and surrounding area, Fireside has never made it big elsewhere. That is a pity. At the Minnesota Agricultural Experiment Station, superintendent Charles Haralson crossed McIntosh and Longfield (an older Russian apple that at one time was considered a premier dessert apple). The tree first fruited in 1927, and Fireside was introduced in 1943. It proved to be exceptionally hardy and of superlative quality.

Fireside is a large apple, round to oblate, but often tending to conic. It has a greenish-yellow background that gradually develops deep red striping and blushes of red. With maturity it develops a thick waxy bloom that gives a greasy feel to the skin. A deeper red color sport named Connell Red was found in 1956 in the Minnesota orchard of Tom Connell. It is virtually identical in flavor, but its skin is solid red.

Fireside's flesh is a perfect blend of tart and sweet, with the accent on sweet. It can stand tall against any other apple from warmer climes, and it keeps well into the late winter.

The tree is moderately vigorous and forms a rounded tree. It is annual, productive and hardy up to Zone 3. It is a late-season apple, so be sure you have enough heat units to produce ripe fruit. ■

Freedom

Hardiness Zone 4
Introduced 1958
Origin Geneva, New York
Primary uses Cooking, dessert, baking

Unlike the wilding tree that casts its branches in the path of someone willing to try its hanging fruits for a lark and is *found*, the engineered tree's history can be complex but traceable, often back several generations. Freedom is such a tree, a product of the fruit breeder's science and art.

Most apples have two parents. The history of one of Freedom's parents begins with a now-famous hybrid between the scab-resistant floribunda crabapple (*Malus floribunda*) and Rome, a large red apple discovered in Ohio. The resulting seedling was bred in Illinois by Dr. C.S. Crandall, who labeled it *Malus floribunda* 821.

This seedling was crossed with itself, which produced a seedling that showed promise and was given the adorable name of F2-26829-2-2. This seedling was then crossed with Golden Delicious. One of the progeny, NY 49821-46, was chosen as a parent for a fourth generation.

Freedom's second parent was also a seedling, but one that involved named cultivars. Its parents were Macoun and Antonovka, a Russian apple of good quality and unrivaled hardiness that had been identified as being resistant to a different strain of apple scab. It was hoped that introducing these genes into the breeding line would bring resistance to both strains of apple scab. The chosen seedling of this union was named NY 18492.

The union of these two parents, NY 49821-46 and NY 18492, produced Freedom, which, as they had hoped, had resistance to both strains of apple scab. Now wasn't that simple?

Freedom's large symmetrical fruit is oblate with a yellow background, becoming nearly covered in bright red stripes and blushing in full sun. Prominent white lenticels are spread evenly across the skin. The cream-white flesh is medium-fine in texture, juicy and sprightly subacid.

This apple has never received the attention it deserves. It is a very useful multipurpose apple, being a fine dessert apple, as well as an excellent baking and juice apple.

Freedom is a tree of great vigor. The new growth is thick with a red-brown bark that is shiny toward the base, becoming tomentose (woolly) at the tips. Like the fruit, it has noticeable white lenticels scattered over the bark's surface. It is one of the most distinctive scion woods for the grafter. The branching structure is relatively upright, becoming more spreading as the tree ages and begins fruiting heavily. And fruit heavily it does, with dependable annual crops. It also begins fruiting at a young age.

This is an ideal apple for the organic grower looking for a heavy-yielding tree with excellent quality and great attributes for processing into a wide range of products. ■

Frostbite

Hardiness Zone 3
Introduced 2008
Origin Minneapolis, Minnesota
Primary uses Dessert, cooking

The most intense experience I have had with an apple was my first bite into a Minnesota 447. I had just begun growing apples and was discouraged by the Red Delicious and Jonathan trees I had purchased at a local chain store. They suffered partially from my lack of knowledge but more from extreme winter temperatures and a short season that did not allow them to mature properly.

My second purchase was from Boughen Nurseries in Manitoba. This nursery grew apples for extreme northern conditions. One of the offerings was an unnamed seedling from the University of Minnesota breeding program called Minnesota 447.

I had never heard of selling apples with numbers instead of names, but I was intrigued by its description in the nursery catalog. It took several years for the tree to bear on my dry, gravelly hillside, but when that first apple broke apart in my mouth I tasted a different world of apples — much headier than any I had experienced before.

I later moved and started a nursery. After many years I returned to my old home and collected scions from Minnesota 447. I grafted them onto Ottawa 3, a dwarfing rootstock, and planted them on another dry hillside. Two years later I was crushed when what I believed were Minnesota 447 trees had been run over by a tractor.

Some years later, strolling through the orchard, I spied two small trees that each had several fruits. They did not look familiar, and I reasoned they might be Bethel, an old apple whose skins are also deeply colored and spotted. When I bit into the fruit I was astounded.

"Taste these Bethels," I called out to my retail manager, Bruce.

"This isn't a Bethel," Bruce replied. Suddenly I knew. The taste of that first Minnesota 447, eaten so many years before, flooded back into memory. The trees had not been crushed by the tractor. I was ecstatic.

Describing the flavor of this apple is challenging, as if the complex essences of this apple have been distilled and concentrated. It is like comparing wine to a well-aged brandy. The fruit has a low-key tang but with a sweetness some have compared to sugar cane or molasses. Others claim they taste pineapple. Additionally it has an underlying taste of nuts and an incredible crispy crunch when you bite into it.

Minnesota 447 was used as a parent in the breeding of Keepsake, Sweet Sixteen and Zestar!. Recently the parentage of Honeycrisp, the current superstar of hardy apples, was confirmed by DNA testing, and it turns out its parents were not Macoun and Honeygold, as believed, but Keepsake and an unknown cultivar. That means Minnesota 447 — this strange, small, wonderful, intense apple — is a grandparent of the spectacular Honeycrisp. It all makes sense.

I was not the only fan of Minnesota 447. For 30 years those who had tasted it badgered the university to name it. A contest was started and

over 7,000 entries from around the world were received. Eight people sent in the name chosen.

We now have a "new" apple: Frostbite.

The naming of this small, humble, oddly colored apple with a wildly different flavor is a testament to its uniqueness and to the persistence of the people who knew that this apple could not be left to slide into horticultural oblivion. It was a great event for palates everywhere. Though I abhor naming favorites, count this apple as my favorite. ■

Gala

Hardiness Zone 4
Introduced 1974
Origin New Zealand
Primary uses Dessert, cooking

In the 1920s the New Zealand orchardist and breeder J.H. Kidd crossed his new seedling Kidd's Orange Red (a cross of Delicious and Cox's Orange) with Golden Delicious. Gala, the result of this marriage, has risen to become one of the most widely grown cultivars in the world. It is adaptable to a wide range of growing areas and has proven to be a mainstay of the supermarket and the local farmers market.

Gala is a medium-sized apple that is oval in form with noticeable knobs at the basin. Its skin is a soft to golden yellow with numerous red and wispy pink stripes decorating the surface. Since its introduction there have been many color sports — far too many to mention. Royal Gala is perhaps the best known. Many claim these more highly colored sports are not as tasty as the original, though some believe Imperial Gala has a flavor truest to the original.

The tree is vigorous with a spur-type habit. It produces copious annual crops and begins producing at an early age. The fruit is virtually free of blemishes and does not bruise easily, a welcome trait for the commercial trade. When fresh off the tree, Gala's flesh is crisp and juicy with a well-balanced acidity and fruity highlights. This cultivar requires several pickings to ensure it is picked at optimal ripeness. If picked early or kept on the tree too long the taste will become duller.

The market loves Gala because it is reliable, ships easily, is appealing to the eye and stores well. Because it is produced in both the northern and southern hemispheres, it can be found in virtually every store year-round.

This apple is being grown in Zone 4 with good success. In colder areas, even if the tree survives, its fruit will not properly ripen.

Although many pooh-pooh Gala, saying it is "too common," you must remember that in horticulture "common" means it is adaptable to many places and acceptable to many peoples. In short, "common" is a compliment. It very well might be the most popular apple in the world today. You may prefer edgier apples, but you cannot deny this undisputed heavyweight in the apple world its due. ■

Ginger Gold

Hardiness Zone 4
Introduced 1989
Origin Nelson County, Virginia
Primary uses: Dessert, cooking

In some respects this apple can be considered the child of a hurricane — in this case Hurricane Camille, a vicious storm that devastated Nelson County, Virginia, in 1969. The orchard of Ginger and Clyde Harvey was hit particularly hard. As Clyde was trying to replant part of a Winesap orchard that had been severely eroded by flooding, he noticed a different-looking seedling among the tangled and uprooted trees. Luckily for us he planted it with the others. The interloper began producing yellow apples that closely resembled what is now assumed to be one of its parents, a Golden Delicious that grew nearby. The other parent has been tentatively identified as Albemarle Pippin (a strain of Newtown Pippin). Clyde soon realized the value of this upstart and named it after his wife, Ginger — not for its flavor as many believe. This unusual beginning has proven a huge boon to the apple world. It was introduced into the commercial market in 1989.

The fruit begins grass-green and changes to soft yellow, sometimes with a reddish blush, and it closely resembles the shape of Golden Delicious. Its creamy white flesh is sweet for an early apple but has more tartness than its famous parent. Ginger Gold can be kept in refrigeration for several weeks with little loss in quality. It is most popular as a dessert apple but is used for sauce and salad, as it does not readily brown. Many have rated this newcomer as one of the best early apples introduced in years.

The tree is a vigorous grower that begins bearing early in life and produces good crops annually, particularly if thinned, which will help increase the size of the fruit. The fruit ripens in mid-September in New Brunswick. Ginger Gold is generally healthy, though it is quite susceptible to mildew and fireblight, particularly when young and growing quickly. Its hardiness is still being tested in the north. It will probably not be hardy in areas colder than Zone 4.

This is an excellent addition to the stable of high-quality early apples. Though still a bit new and relatively untested in colder climes, it is worth planting at least one to savor its special flavor when the hunger for a fresh apple is at its peak. This cultivar has also provided us with yet another great name and fascinating story from the annals of horticultural history. ■

Golden Delicious

Hardiness Zone 4
Introduced 1916
Origin Clay County, West Virginia
Primary uses Dessert, cooking

Golden Delicious grown in the north have more pink blush and are often slightly russeted.

I have a single tree of Golden Delicious in my orchard. It is actually a sport called Smoothee that was chosen for its reduced tendency to form russet on the skin, which apparently growers abhor. I cannot fault this tree for hardiness. Despite temperatures that have dipped to –40°F (–40°C), it has never had winter injury. I believe it helps that it is grafted on Beautiful Arcade rootstock, which is extremely hardy. The problem at my site is the lack of sufficient heat to properly mature the fruit. Although quite edible by the very end of the season, my Golden Delicious lacks the perfumed sweetness attained in warmer regions and does not have the creamy yellow skin it is famous for. Instead the fruit is yellow, spotted red and dappled with bits of russet.

This apple has one of the better-known histories of any. It was found as a wild seedling on a hillside in Clay County, West Virginia, in 1890 by the young son of a farmer named L.L. Mullins. The boy, J.M. Mullins, was scything the field when he came across a small apple seedling. As there were few in the area, he let it live. Oh, how close it came that day to never becoming the most famous apple in the world. The seedling grew into a sizable tree and began producing copious crops of large and spectacular-looking apples. The farm was bought by the boy's uncle, Anderson Mullins, who noticed the quality of the fruit. Not only was it better than the nearby Grime's Golden (a probable parent), but also it had superb flavor, it kept through the winter and the tree was very healthy.

Stark Bro's Nurseries & Orchards Co. in Missouri, the largest fruit tree nursery in North America at the time, held yearly contests to discover new cultivars. Mullins decided to submit some fruit of the tree he had named Mullins Yellow Seedling to the contest. After Mullins sent a second batch of fruit in April to show how well it kept, Stark Bro's realized the value of the seedling. Paul Stark took a train, then rode by horse to the farm that fall. No one was home so he wandered about the farm till he came across a tree that outshone all the others in the area. He knew it had to be Mullins's seedling. He bit into one of the apples and was said to have exclaimed, "Eureka, I have found it!" Perhaps this quote was created for effect, but the gist is certainly true.

Stark paid an astounding $5,000 for the rights to the tree and the ground around it. He erected a cage of posts and wires so that no one else could take scions from it. In 1916 it was

The tree is wide and spreading yet very strong. It produces a large annual crop.

Though quite hardy, Golden Delicious will not properly ripen in areas colder than Zone 4.

introduced as Golden Delicious, a nod to the Red Delicious apple Stark Bro's had purchased from a farmer in Peru, Iowa, which had become their best-selling apple.

This is an apple for the lovers of sweet, though there is just enough acidity to keep the balance from tipping into the dark abyss of blandness. It breaks easily in the mouth, being firm yet melting, is somewhat spicy and has enough juice to satisfy. Golden Delicious has legions of worshippers, including many breeders who have used it to create an immense stable of apples that owe a portion of their genes to this cultivar. At this moment it is, if not the most important apple in the world, certainly among the top three.

For us in the north, this is a worthwhile apple, but ours will be firmer, less aromatic and with more of that russet that growers in places like the Wenatchee Valley, Washington, view with alarm. ■

Golden Russet

Hardiness Zone 4
Introduced 17th or 18th century
Origin England or New Jersey
Primary uses Dessert, cider

Just as there are people whose names bring hushed reverence, apples like Golden Russet are spoken of with the deference given to an ancient sage. Some believe this apple has its origin in England sometime around the late 17th century, though many trace it to Burlington County, New Jersey, sometime in the late 18th century. It is unlikely we will ever know the truth. As with most apples at the time, its initial fame was due to its qualities as a superb cider apple, and we are not talking sweet cider here. Without refrigeration even hard keeping apples had a limited storage life. Most were pressed for juice and fermented into hard cider, a product that was always in demand, kept indefinitely and gave a far better return to the farmer than a barrel of pulpy fruit.

The later history of Golden Russet becomes a bit muddied, as there were several seedlings, most likely offspring of the original that became known as golden russets. The original is a small to medium-sized round oblate fruit. Its name refers to the leather-colored russet skin, which has a feel somewhat like chamois leather. There is an underlying green tone, and when fully ripe, a slight, dull reddish cheek can develop on sun-drenched specimens. Those that show more red are likely not the true cultivar.

This is a late apple, picked in early November in New Brunswick, and great care must be taken to harvest when fully mature. Pick too early and the fruit will carry a "green" taste, even when

The classic and best-known russet apple.

stored for several months. Picked properly, the fruit will develop complex flavors married to a restrained sweetness. It should be stressed that this is not an apple to eat at harvest. Though palatable, the fullness of its flavor will not develop until stored for a month or two. Proper humidity is also essential as dry storage will cause it to wither.

In the northern hemisphere this apple is usually offered at Christmas or in early January. One of the distinctive features of Golden Russet is its fine-grained texture. A perfect fruit will crack apart when bitten and, though not as juicy as some, its flesh will melt in the mouth like cream.

Golden Russet is surprisingly hardy, handling temperatures approaching –40°F (–40°C). The limiting factor is more often the length of season. Because it is so late in ripening, areas where severe freezing occurs before it is ripe cannot grow this apple to perfection. The tree itself is spreading with a willowy habit, and

Golden Russet trees tend to be at least as wide as tall, although this tree has lost upper branches.

Fruit is evenly spaced throughout the tree and is often borne on the tips of branches.

often fruit can be seen on the tips of those delicate stems, for it is one of the few apple trees that produces fruit buds on the current season's growth, making pruning and picking more difficult than trees that produce fruit on spurs.

The new branches are more slender than those of other apples, and they are a reddish brown with prominent lenticels. It is wise for northern growers to find a hardy dwarfing rootstock, for this cultivar can take six years or more to come into production on seedling roots. The tree is long lived and healthy. Though not immune to apple scab, it is rarely a problem. No doubt the rough texture of the fruit's skin is not as hospitable a surface for scab spores to germinate.

Golden Russet has always maintained a loyal fan base, but because the fruit is too small for most growers, the number of trees grown has declined over the past half century. However, a new fan base is building. Cider makers are increasing in numbers, and this apple is one of the most sought after. It is a premier apple for hard cider. If you have a hankering to discover the subtleties of the art, Golden Russet should be put in the mix. Many of us, however, will be happy biting into this most venerated of apples on a crisp winter's day to experience once again its unique assemblage of sweet and subtle aromas. ■

Goodland

Hardiness Zone 3
Introduced 1955
Origin Morden, Manitoba
Primary uses Dessert, cooking

The search for good-quality apples that can grow in the northern regions of Canada has been a long and often frustrating ordeal. Starting in 1888 with the establishment of the Dominion Experimental Station in Brandon, Manitoba, researchers and amateur fruit breeders grew thousands of seedlings in hopes of finding the golden child that would allow the most cherished of fruits to grow where winter temperatures could hover between –40°F and –58°F (–40°C and –50°C).

Using Russian apples originally brought in by Professor Budd of Ames, Iowa, crosses were made with the aim of creating hardy, good-quality fruit. Their efforts generally failed. A seedling might show great promise for several years until killer winters, such as those that occurred in 1915–16 and 1942–43, arrive to wipe out any progress.

The search was joined by breeders like Arthur Coutts and Adolph Heyer of Saskatchewan, who created apples, such as Heyer 12, that held some promise. However, the hundreds of apples raised by amateurs, usually with unregistered names, created quite a bit of confusion.

Eventually the combined efforts of the Morden Research Station, the University of Saskatchewan and the University of Alberta began to yield some results. One of the most promising was a seedling of Patten Greening. The tree, which yielded a large green apple with a rosy blush, was selected in 1925 and named Goodland in 1955. While not hardy enough for the very farthest regions of the north, it survives throughout Zone 3 with little or no damage.

Its creamy white flesh is fine and crisp with more aromatics than most of the super-hardy cultivars. It is good for both dessert and cooking, and keeps for two months, far longer than most of the toughest apples.

The tree is vigorous, upright and self-thinning, creating good-sized fruit every year. In the prairies it ripens in early October. Farther south it is ready by mid-September.

Breeding work continues, and new cultivars suited for the far north are being introduced that may eventually eclipse this pioneer apple. Nevertheless Goodland is still one that those in the north can depend on, and its continued availability is a testament to its value. ■

Granite Beauty

Hardiness Zone 3
Introduced ca. 1815
Origin Maine and New Hampshire
Primary use Dessert

It would be hard to imagine a more unbelievable story for an apple than the tale of Granite Beauty. The story begins in the early 1800s when a lady named Dorcas Dow visited her friends in Kittery, Maine. When leaving she requested a riding crop. A small apple seedling was pulled from the side of the road and handed to her. Once the horse had returned her to her pioneer home in Weare, New Hampshire (then called Halestown), she promptly planted the apple seedling, as orchards were an important part of establishing a homestead. Once it began producing apples, Dorcas laid claim to it, and soon the surrounding area knew it as the Aunt Dorcas apple. Later it came to be known as the Grandmother apple or the Clothesyard apple.

Now called Granite Beauty, after the Granite State (New Hampshire), this apple was spread to nearby locations around New England. It never found widespread notoriety, but it has lately been brought to the attention of the apple world by the seekers of historical cultivars who haunt the old orchards around them.

The tree is very hardy and bears good crops of pale-yellow fruits that become striped, then nearly completely blushed with dull iron red. The skin has a greasy feel, which may partially account for its good keeping ability. It has been esteemed as a wonderful dessert apple. Its flesh is cream white, tender, crisp and juicy. Some pommeliers have detected notes of cardamom and coriander. The stem is thick, of medium length and set in a deep cavity. The basin is deep and wide.

It is exhilarating and somewhat curious that we are just now tasting fruit that others long ago picked and laid into their root cellars for the coming winter — fruit from apple trees that nearly went the way of the dinosaur and dodo. And to think, some are improbable beauties that grew from a riding crop. ■

Greensleeves

Hardiness Zone 4
Introduced 1966
Origin Kent, England
Primary uses Dessert, cooking

I sing the praises of a relatively recent selection from England, which has remarkable hardiness when grown on hardy roots. Greensleeves seems to have never "made it" in the markets of North America, but it is a reliable apple where temperatures can drop to –40°F (–40°C). Quality, however, is its lasting legacy. A blend of tart restrained by sweet is perfectly married to a crisp, crackling texture. The flesh is melting, and the skin never gets in the way of enjoyment.

This apple originated in Kent, England, as a cross between Golden Delicious and James Grieve and was introduced in 1966. It has the sweetness of Golden Delicious but with a burst of tartness inherited from its other parent, which is an old Scottish cooking apple. In 1993 Greensleeves won the Royal Horticultural Society's Award of Merit. I know some have reported it as unexciting in taste, but it is my belief that these reports are from more southerly sites. I think this apple is most at home in the north, where its flavor reaches its apex.

The tree has a great form that pruners will love. Well-angled branches are strong, and they need to be. Greensleeves is precocious and a heavy annual bearer of good-sized fruit. The glossy leaves are a deep lime green. Scab has never been a problem on either leaves or fruit. The incidence is minimal most years. Most importantly, it seems extremely resistant to apple maggot.

The fruit is uniformly round to slightly oblate, with a smooth and uniform green color sporting a delicate pattern of evenly spaced ivory lenticels. At maturity the skin becomes a soft butter yellow. Given proper storage and harvesting time (early October in New Brunswick), Greensleeves will keep well. Unpicked fruit will hang on the tree into the winter.

Greensleeves certainly is one of my picks for the northern organic orchard of my dreams. I have been in love since my first bite. ■

Haralson

Hardiness Zone 3
Introduced 1922
Origin Morristown, Minnesota
Primary uses Dessert, cooking

The most up-to-date history of Haralson begins with a man named Seth Kenny of Morristown, Minnesota. He planted the seed in 1908 and selected the seedling in 1913. After distributing it to other growers for trial it was named after Charles Haralson, the superintendent of the Minnesota Agricultural Experiment Station. Until now it has been the most planted apple in that state's orchards, though by the time you read this Honeycrisp and Zestar! will likely have changed things. In the past its origin was listed as "Malinda x Ben Davis," but DNA testing unveiled that the pollen parent was Wealthy.

Haralson is a superbly hardy apple, an obvious prerequisite for this northern state. It is a relatively small-statured tree with a strong central leader and wide-angled branches. Bearing is somewhat biennial, so thinning is recommended to produce more even cropping.

This is an apple with a tangy tartness, though allowing the fruit to fully mature on the tree will provide more sugar. Though most often used as a dessert fruit, it creates wonderful pies because of its high acidity.

The fruit is round to slightly conic and exhibits an even red striping over the entire surface. Several deeper red sports are available. These are more or less entirely red, showing little striping. They include Red Haralson, Haralred and Scarlet O'Haralson.

For those living where spit can freeze before it hits the ground, this is a good choice. ■

Honeycrisp

Hardiness Zone 3
Introduced 1991
Origin Minneapolis, Minnesota
Primary uses Dessert, cooking

No other recent introduction has had as much impact on northern fruit growing as Honeycrisp. Its near-instant acceptance as the apple of choice for millions of consumers is unprecedented and has probably saved hundreds of northern orchards from bankruptcy, as intense competition from apples unsuited for northern growing destroyed what little profit these growers had come to expect for their labors.

Honeycrisp originated from the breeding program at the University of Minnesota. Its parents had been listed as "Macoun x Honeygold," but recent DNA testing has revealed that one of its parents is Keepsake, an older Minnesota apple with a unique nut-like flavor. The other parent does not match any known cultivar and thus remains a mystery.

The fruit is medium to large when well thinned. It is roundly oblate to conic. The skin has a yellow background that becomes overspread with dull red striping and eventually blushed red over half its surface. The flavor is well balanced twixt tart and sweet, though not particularly complex or aromatic. It is the crackling crunch and juice that sends people into paroxysms of delight. This crispness holds even after long storage periods. In normal refrigeration with no atmospheric manipulation, Honeycrisp will remain crackling for six months. In controlled atmosphere storage it can remain so for nearly a year. The result is a fruit grower's dream: an apple with good flavor, superb texture and a long shelf life.

The tree is only moderately vigorous and slightly spreading. It ripens in midseason (late September/early October in New Brunswick). It is very early bearing and is dependably annual, though its tendency to bear heavily each year can eventually have consequences

for the grower. Research has shown that the trees need to be spur pruned on a regular basis to maintain a good balance between vegetative growth and fruit production. This involves thinning the spurs where flowers are produced. By removing one-third to one-half of the spurs each year the tree is able to grow more new wood for future spur production, and fruit size is kept larger with better quality. Some growers are choosing to graft Honeycrisp onto a more vigorous root to create a better-sized tree initially.

The apple world would be a poorer place without this sensational new introduction. Perhaps the name of Honeycrisp's true parent can be applied to its character — a real keepsake. ■

Honeygold

Hardiness Zone 3
Introduced 1970
Origin Minneapolis, Minnesota
Primary uses Dessert, cooking, cider

Our orchard has grown Honeygold for nearly 40 years, acquiring it only 12 years after its introduction in 1970. This is another achievement of the University of Minnesota breeding program, which is certainly one of the most prolific in the world. Honeygold is a cross of Golden Delicious and Haralson. The goal of creating a sweet Golden Delicious–type apple with the hardiness of Haralson was certainly achieved with this apple.

The fruit is medium to large and irregularly round to conic. Its yellow skin develops a bronze-red blush on the sunny side. The crispy yellow flesh is sweet with good juice, similar to its parent Golden Delicious.

The tree is listed as being hardy to at least –40°F (–40°C) or colder, though the areas with the shortest season might have trouble maturing the fruit to peak flavor. Suffice to say, this tree is tough. It bears annually and heavily, and the young trees are precocious. What's not to like?

Although it is not generally known as a cider apple, Claude Jolicoeur of Quebec has used Honeygold for cider and finds it high in sugars with an ideal level of acids. He uses it for his late-season blends.

When properly ripened Honeygold is a superb Golden Delicious–flavored apple.

The tree is hardier than its parent, Golden Delicious.

Well, the problem we have had is with apple scab. Though touted as resistant, our experience in our admittedly humid environment is that it is quite susceptible to this ubiquitous fungal disease. Without protection much of the fruit is unusable. In areas with drier growing seasons, this may not be as much of a problem. If you can provide protection from scab, the fruit will satisfy most every palate. Some have said Honeygold is a bit blander than its parent, but you cannot please everyone. I know that local growers here have no problem selling their crops. ■

Honeygold ripens several weeks ahead of Golden Delicious.

Hudson's Golden Gem

Hardiness Zone 4
Introduced ca. 1940
Origin Tangent, Oregon
Primary uses Dessert, cooking

Not many russet apples have been introduced in the past century. Though they have their fans, some will not try a russet apple because of its leathery feel and brown skin. For those of you like me who adore russets, here is one that is getting rave reviews for flavor. We have just planted a couple in the orchard, so time will tell how adapted it will be for our site.

Hudson's Golden Gem was found as a chance seedling along a fencerow at the Hudson Nursery in Tangent, Oregon, around 1930. It is delightful to discover that Mr. Hudson wanted anyone with an interest to propagate this tree, as he considered it the finest-quality apple he knew. It is rare these days to find that kind of generosity in the modern horticultural industry, which is a pity.

This cultivar has had little trialing in colder growing sites to my knowledge. The tree is quite thrifty but does not attain a great size, even when grafted on seedling roots. I have found references that list the plant as a Zone 3 hardy tree, but I reserve judgment on this. The tree is productive and annual, though with some biennial tendency. Thinning should solve this problem.

Hudson's Golden Gem is the largest russet apple on the market. It is round to conic, sometimes long conic. The skin is yellow and more or less completely covered in russet. The flavor is touted as among the highest quality of any apple. It is crisp, tender, juicy and sweet with a nuttiness that has captured the hearts of many. In the early years of production, the fruits are susceptible to cracking, but this disappears as the tree ages.

We are hopeful Hudson's Golden Gem will prove of value in our area. It requires very little chilling to come out of dormancy. This is good for southern areas, but it makes me wonder if it might be a problem where winters are cold, and where temperatures can fluctuate wildly as they do in New Brunswick. Only time will tell, but let's keep our fingers crossed. ■

Hyslop Crab

Hardiness Zone 3
Introduced 19th century
Origin Massachusetts and Wisconsin
Primary uses Cooking, cider

In my orchard is a 40-year-old specimen of this wonderful crabapple. It is thought to have originated in Massachusetts but was introduced by a Milwaukee nurseryman named Thomas Hislop sometime before 1869. Later its name was standardized as Hyslop.

This is a large crab with a symmetrical round oblate form. The skin starts as a soft yellow but matures to a deep purple-carmine red with a bluish bloom and scattered pale white lenticels. The flesh is yellow and firm, occasionally showing red streaking. The fruit is tart and juicy, becoming drier as it matures.

If picked at the proper time Hyslop is one of the finest crabs for pickling and jelly. Its good size makes a jar fill quickly and its reasonably long stems make it an ideal candidate for a classic pickled crab that can be picked up by its stem.

The tree is moderate in size with an open spreading form. The twigs are slender and olive green with reddish shading. If you are seeking good grafting wood, you should consider pruning it fairly rigorously to produce heavier scion wood. Crabapples as a rule have thin wood that makes grafting more of a challenge.

This is one crab that, though an older cultivar, should be propagated for the future. It is a healthy annual producer of quality fruit that is unsurpassed for the kitchen. ■

Fully ripe, Hyslop crabs are burgundy red. If left too long they turn mealy quickly.

The tree is open and a pleasure to prune.

Fruit that is not fully ripe is ideal for jelly and pickled crabs.

Keepsake

Hardiness Zone 3
Introduced 1979
Origin Minneapolis, Minnesota
Primary uses Dessert, cooking

I was immediately fascinated by the release of Keepsake from the University of Minnesota. Although its parentage is usually listed as "Malinda x Northern Spy," Keepsake is actually the child of Frostbite. It was Frostbite (originally called Minnesota 447), an open pollinated seedling of Malinda, that was crossed with Northern Spy to produce Keepsake. The cross was made in 1936 by Dr. W.H. Alderman and released in 1979.

Keepsake has many of the same qualities as Frostbite, though it is larger and perhaps slightly more presentable for our demanding modern public. It, too, is crackling crisp when freshly picked with fine ivory flesh. Its other parent, Northern Spy, is famous for its outstanding quality, and Keepsake has that apple's ability to store through the winter. Though it has never become an important cultivar in the larger growing areas, it certainly is grown in Minnesota and has a fan club scattered across the continent.

The apple is medium sized and conic with angular sides. The skin of a sun-drenched fruit is a dull red with occasional rough areas that are referred to as scarf skin, or, as it is referred to in the trade, "bridal veil." This is a physiological condition that occurs on some cultivars when the cells are disrupted and refract light differently. It does not affect taste. The flesh is crisp and juicy, and the taste is sweet and aromatic. The flavor improves with storage. It can be kept until the following spring with little deterioration.

The tree is smaller than most. It is hardy for all but the most extreme climates, although it is fairly late to ripen (mid-October in New Brunswick) so those living in areas with very short seasons may not be able to grow it to perfection. The tree and fruit are somewhat resistant to apple scab, fireblight and cedar apple rust.

Keepsake is another reason to give homage to the tireless efforts of the breeders at the University of Minnesota. They have given us many of the finest hardy apples. ■

Liberty

Hardiness Zone 4 (possibly 3)
Introduced 1978
Origin Geneva, New York
Primary uses Dessert, cooking, cider

Here is yet another fantastic introduction from Cornell University and its associate, the New York State Agricultural Experiment Station, an institution famous for its prodigious output of horticultural information and outstanding fruit cultivars. This is the origin of such famous apples as Cortland, Empire, Jonagold, Macoun, Spigold and, now, Liberty.

Liberty is the result of a cross between Macoun, a smallish apple with gourmet flavor, and a scab-resistant seedling called PRI 54-12. While Macoun is highly susceptible to scab, its progeny Liberty is resistant to a host of diseases, including apple scab, cedar apple rust and fireblight; however, Liberty's flavor vies with Macoun for excellence.

The medium to large fruit is round to conic. The yellow-green background takes on a deep, rich red where sun is abundant in the tree. The stem is short and stout. The flesh is crackling crisp, white with some pink veining and somewhat aromatic and refreshing.

Careful timing is essential with Liberty. Pick too early and it will lack flavor; pick too late and it will lack crispness and storability. If picked at the proper time (early October in New Brunswick), it will thrill the taste buds and keep well into winter under good storage conditions. If for some reason the pickers never arrive, Liberty can hang onto the tree until well into winter. Strangely enough, Warren Manhart in his book *Apples for the 21st Century* (1995) claimed that he had to pick his as soon as the first drops occurred, as after that they all fell. His orchard is in Oregon. If not picked, mine hang onto the tree till nearly spring. It is fascinating how apples behave so differently in different locations. Perhaps in the future someone will create an ice cider with these frozen orbs.

Liberty makes a wonderful sweet cider with a near-syrupy consistency. It is an excellent bulk juice for hard cider as well, providing great body and flavor.

Liberty is ideally suited to organic production. It makes a fantastic sweet cider with great body.

The tree is strong and well built.

The tree is exceptionally hardy and will do well in Zone 4 and possibly in protected areas of Zone 3. It has a sturdy and well-proportioned frame with strong crotches. Fruit is produced evenly throughout the tree. It bears prolifically every year, so growers should consider thinning to maximize size and quality. The leaves are deep green and glossy.

This is certainly one of the most important introductions from the new generation of disease-resistant cultivars and deserves a place in any northern orchard. For those considering organic apple production, this should be one of the first choices. It comes as close to flawless as I have observed. ■

Liberty is annual and very productive. It's among the finest cultivars we have grown.

Lobo

Hardiness Zone 4
Introduced 1930
Origin Ottawa, Ontario
Primary uses Dessert, cooking

There have been a host of apples bred with McIntosh. Lobo is one of the finest of its progeny. This was an early cross, bred in Ottawa, Ontario. The seed was planted in 1898, first fruited in 1906 and then released in 1930. Although it is generally regarded as a seedling of McIntosh, and it certainly has many similar characteristics, it was also reported to be a seedling of Langford Beauty. Perhaps Langford Beauty was simply the other parent. Another horticultural mystery that could probably be solved by DNA analysis.

Lobo is grown principally in Eastern Canada and, until recently, was a major cultivar in the Maritime provinces. It can be found in a few places in Europe as well. It is slipping in popularity as newer cultivars steal the spotlight, and the world gives short shrift to the older standbys.

The fruit is distinctive, having a somewhat squat pumpkin-like shape and a deep red color to the skin with distinctive white dots (lenticels). The flesh is similar to McIntosh but with perhaps slightly more acidity, which makes it a great pie and sauce apple. However, it is best known as a dessert fruit.

Lobo ripens two weeks ahead of McIntosh and fills a niche for those who cannot wait for their first Mac. It is less susceptible to scab than its parent, but if you desire clean fruit, you will have to adopt preventative measures.

This is another lovely apple that is gradually slipping into the assisted living quarters for retiring apples. ■

Lodi

Hardiness Zone 3
Introduced 1924
Origin Geneva, New York
Primary use Cooking

If you visit Lodi, New York, you might find a few of these apple trees growing around the area. They have been there since 1924, when Lodi was introduced as an annual Yellow Transparent–type apple. The cross between Montgomery and Yellow Transparent was first made in 1911 at the New York State Agricultural Experiment Station, not too far from Lodi. The famous Transparent apple was notorious for bearing only every other year. Lodi was to replace it as an apple that had similar qualities but was reliably annual. In this it only half hit its target, as Lodi is also somewhat biennial, but off years do have some crop.

This apple has a distinctive form, being round conic and very symmetrical. The stem is long and stout. The skin is yellowish green, and the lenticels are scattered dark dots. Light russeting often occurs in the stem cavity and occasionally as random streaking.

The white to slightly greenish flesh is tender and juicy, like its parent. Its acidity level makes it a good cooking apple, though it is quite acceptable as a dessert apple when picked at the proper time. A great advantage over Yellow Transparent is Lodi's firmer core and slightly longer shelf life. For many years this apple has been used as an indicator of the ripening period for an early apple.

Lodi blossoms in the early to midseason and ripens in midsummer (early August in New Brunswick). The tree is vigorous, and the branches have an upright curve to them unless weighed down by a heavy crop. It is quite winter hardy, but the caveat is that it tends to go slowly into dormancy and a quick descent into winter temperatures can cause damage. ■

Macoun

Hardiness Zone 4
Introduced 1923
Origin Geneva, New York
Primary uses Dessert, cooking

The tree is smallish and moderately productive.

If the gourmands of the apple world were to convene to decide which apple would win the prize for Gourmet Apple of the 20th Century, Macoun would be a contender. A quick browse through comments made about this cultivar shows a fanatical loyalty by its fans. Having tasted this off my trees, I concur. Despite this rabid response, Macoun is relatively unknown outside of its place of origin — New York — and the surrounding area, perhaps because it seems to grow to perfection only in these areas.

Macoun began life as a seedling in a breeding program of the New York State Agricultural Experiment Station and was introduced in 1923. A cross between McIntosh and Jersey Black, the fruit has a dense, crisp, pure-white flesh and a deep red skin with purple tones. The round to somewhat conic fruit is perfect for snacking, but it is not appreciated by growers who like large apples that increase their pack-out. Macoun does not keep well into the winter, so it is only available in October and November, another reason this apple will only remain a cultivar for the discerning grower.

The tree is smallish with an open spur–type growth habit. It becomes rounded with age. It has proven hardy in Zone 4 and might be worth trying in 3b in areas that are protected from winds and late frosts.

I am somewhat sheepish about including Macoun as it is very susceptible to apple scab, but it is such a marvelous eating apple that it should be considered by those looking for excellence in quality. ■

Though it ripens late and does not keep through winter, Macoun has a flavor that is second to few.

Mann

Hardiness Zone 3
Introduced 1850
Origin Oswego County, New York
Primary use Cooking

There are certain apples that never found their way into the orchards of the modern era. Mann is a great example. This cultivar was a chance seedling discovered by one Judge Mooney in Oswego County, New York. The apple was later introduced into Niagara County, New York, by a Dr. Mann and was named for him by the Western New York State Horticultural Society. Mann has also been known as Diltz or Deiltz.

This apple is oblate in form, medium in size and the deepest forest green that morphs to a dull honey yellow at maturity. When it is growing, its white lenticels contrast well against the green for a very appealing look. Russet patches are sometimes seen emanating from the stem cavity. The flesh is yellow, crisp and juicy, and it is one of the last apples to be picked. It will keep in common storage all winter and would no doubt keep till the following summer in controlled storage.

We have grown this apple for nearly 45 years, and it has stood up to anything winter could throw at it, but beyond hardiness its resistance to disease and insects is what makes it stand out. It is among the cleanest apples in the orchard. Mann is somewhat biennial so large crops alternate with smaller crops.

Although it can be eaten fresh, it has always been known as a cooking apple. This is one you could keep in a cold cellar and take out anytime during the winter for that special pie.

This is a large tree on standard roots, so putting it on a semi-dwarf rootstock might keep it at a more convenient height. ■

Mann is a very fine cooking apple with density and great keeping ability.

The tree is vigorous, broad and extremely hardy.

Mann is among the best apples for scab and insect resistance.

McIntosh

Hardiness Zone 3
Introduced Early 19th century
Origin Dundela, Ontario
Primary uses Dessert, cooking

It could be argued that no other apple has had more influence on northern apple growing than McIntosh. For the past century it has been the number-one apple in New England, New York, Ontario, Quebec and Atlantic Canada, and it has been used in many apple breeding programs around the world.

The original tree was found as a seedling by John McIntosh, a young Scot from the Mohawk Valley of New York State. It seems Dolly Irwin, the love of his life, left New York in 1796 with her family to settle in Canada during the American Revolution, as her father was loyal to the English king. Sadly, by the time John arrived to court her, she had died. In 1801 he married Hannah Doran, and 10 years later he traded parcels of land with his brother-in-law and settled on a plot near the village of Dundela, Ontario, not far from the St. Lawrence River. While clearing land on the farm he came across a cluster of about 20 apple seedlings. As apples were still fairly rare in the country, he left them, returning several days later to transplant them to the clearing near the house he had built. Eventually all but one of the seedlings died. The remaining tree proved very hardy and produced large bright-red apples with a wonderful flavor. It has been conjectured that these seedlings were from the leavings of Fameuse apples, an apple popular in the region, though this has yet to be confirmed by DNA testing.

For many years the McIntosh family used the apples for themselves, calling them Granny's Apples. McIntosh tried selling seedlings of the tree, but as trees do not breed true-to-name the seedlings were inferior. In 1835 a traveler taught John and his son Allan how to graft, and from that point on Allan began propagating grafted trees from the original. The fame of this new apple spread throughout the countryside. By the early 20th century McIntosh Red, as it was originally called, had become one of the most important apples in the northeast.

While McIntosh has great flavor, the grower needs to be aware of its propensity for scab.

McIntosh has many excellent qualities. Its flesh is bright white, juicy and melting, crisp when freshly picked with a flavor slightly more sweet than tart, a balance that most people find ideal. When ripened on the tree it takes on a vinous quality. The skin is an attractive bright red that many find appealing. Although the skin is considered on the tender side, if handled carefully the apple can be stored till midwinter or longer in controlled atmosphere storage. As noted, the tree has been hardy through Zone 4 and can be grown in Zone 3, though it may occasionally experience winter injury.

Its flavor will tend to flatten in storage, losing the vinous quality that so perfumes the fresh fruit. Some fruit is also harvested too early and tends to have green flesh that is grassy, not how a properly ripened apple should taste. It does become softer over time as well.

Perhaps its greatest fault is an extreme susceptibility to apple scab. Few apples require as much attention to prevent this disease. Despite the faults, McIntosh remains exceedingly popular today, though many orchards are changing to newer cultivars that are crisper and more disease tolerant.

The tree is a strong grower, eventually forming a rounded tree with slightly pendulous branches. It is a consistent annual producer.

The genes of McIntosh are in many of the most popular apples grown today. The object of many breeders has been to capture the flavor of McIntosh without its disease susceptibility. Its flavor can be detected in such cultivars as Macoun, Spartan, Cortland, Liberty, Lobo, Empire and Novamac, to name a few.

Though the organic grower would be unwise to choose McIntosh, one has to pay homage to this superstar. If you are looking for that McIntosh quality but not the pain and expense of many antifungal sprays, you would be wiser to grow Novamac. For this child of McIntosh, we can thank both its breeder, Dr. David Crowe, and Mr. McIntosh for their foresight and labors. ■

The tree is moderate in size, annual and productive with great hardiness.

Preventative sprays for scab should start just before flowering and continue into the growing season.

Milwaukee

Hardiness Zone 3
Introduced Late 19th century
Origin Milwaukee, Wisconsin
Primary use Cooking

This apple was found growing under a Duchess tree. It was pulled up and planted by George Jeffry of Milwaukee, Wisconsin. I have found little information as to its date of introduction, but it was being planted at the end of the 19th century in the Midwest and New York State and found its way northward to New England and the Maritime provinces of Canada. Until recently several orchards along the Saint John River grew it, where it was much sought after by local piemakers. It is now rare.

The tree is strong and healthy, and when well grown will produce an abundant crop annually. It prefers open soils with good drainage. Milwaukee will thrive in Zone 3 provided there is a long enough season to properly ripen the fruit. What sets Milwaukee apart is its later season and its heightened acidity. These qualities make it a wonderful keeper and a fine cooking apple.

This apple is medium to large and fairly uniform. It is the poster child for an oblate apple. The skin is a shiny pale yellow that becomes blushed with pinkish red and mottled with stripes and patches of carmine-rose. The small white lenticels are quite noticeable. Inside is a white flesh tinged yellow that is crisp, tender and juicy, with a high level of acidity. It is not considered an eating apple, but fans enjoy its citrusy tartness. After all there are many who enjoy sucking on lemons. It is a marvelous cooking apple that will keep into midwinter without losing its acidity. Its texture makes it ideal for slicing and drying. All in all, a chef's delight. ■

New Brunswicker

Hardiness Zone 3
Introduced Late 19th century
Origin Upper Woodstock, New Brunswick
Primary uses Dessert, cooking

An author might be forgiven for choosing an apple so closely tied to his home. This cultivar could never be an important commercial choice today. It does not produce reliably annual crops, tending to a biennial habit, and is perhaps a bit soft for a shipping apple. It is early to midseason and does not keep long. That being said, New Brunswicker is both a historically important apple in the province of New Brunswick and northern New England and a superb hardy, healthy and long-lived apple. It is also a naturally small tree.

The tree bears medium-sized round to oblate fruit with a yellow background that becomes blushed and prominently marked with red striping. The fruit is crisp at maturity but breaks down quickly if not refrigerated immediately. In storage it will keep for a month or so. The flesh is sprightly but aromatic, even before being cut open. A basketful of New Brunswicker will fill a room with the smell of fall. The flesh has been described as white as flour. It is good for fresh eating in season, makes great sauces and pies, and was much esteemed for drying.

Beyond the fact that it is a valuable early apple that can be used for dessert or cooking, it has been the long debate about New Brunswicker's relationship to Duchess of Oldenburg that has most defined the apple. Research done by Daryl Hunter of Keswick Ridge, New Brunswick, discovered that the mix-up occurred when a pomologist from the

New Brunswicker is a very aromatic apple. Its early ripening makes it great for early season pies and crisps.

The tree is smallish and extremely hardy.

Comparison of Duchess on the left and New Brunswicker on the right.

New Brunswicker will blush more heavily on the sunny side.

Iowa Agriculture College named Professor Budd visited the orchard of Francis Peabody Sharp, widely known as the "Father of Fruit Culture" in New Brunswick. This was the original orchard where New Brunswicker was bred and grew, and Budd took scions of the apple, which resembles Duchess when it is still green and unripe. The apple gained the name Duchess of New Brunswick and was subsequently sold to growers and nurseries in the United States. It was eventually labeled Duchess by the Stark Bro's Nurseries & Orchards Co. of Missouri, the largest nursery in North America at the time — thus beginning the confusion between the two cultivars. Although similar, Duchess of Oldenburg and New Brunswicker are distinct. ■

Norland

Hardiness Zone 2
Introduced 1980
Origin Morden, Manitoba
Primary uses Cooking, dessert

A book on hardy apples should include cultivars for the far north. Norland is such an apple, surviving into Zone 2. This cross between Rescue crab and Melba is very early and will go from not quite ripe to overripe in a short time, so proper timing is critical. Once ripe, the fruit will drop, so it is probably best to pick when they are still firm and a bit tart.

The round apples have a background of yellow-green with a soft blush and striping of pale red on sun-drenched specimens. The flesh has a good tart-sweet balance when at peak. They do not keep well, so it is best to use them immediately. They are best used as cooking apples, particularly for sauce.

Norland has definitely found converts where winter temperatures will freeze your nose hairs in seconds. This Morden Research and Development Centre introduction of 1980 may not ever take the world by storm, but it will satisfy the urge for fruit where none could be grown before. ■

Northern Spy

Hardiness Zone 4
Introduced Mid-19th century
Origin East Bloomfield, New York
Primary uses Cooking, dessert

Though others might proclaim Bramley, Rhode Island Greening, Milwaukee or Cortland the finest cooking apple, these well-meaning fans might be drowned out by the cries of those espousing their favorite — Northern Spy, or Spy to those who still grow it. At one time this apple ranked at the top of cultivars growing in places like New York, New England, Nova Scotia and Ontario. To its credit Northern Spy is still being grown to feed processing plants in places like the Annapolis Valley of Nova Scotia. You would be hard put to find any other apple whose hold on its adherents has been so tenacious.

The history of this apple apparently starts with a batch of seeds that was brought from Salisbury, Connecticut, to East Bloomfield, New York, and planted in the seedling orchard of Herman Chapin around 1800. The original tree died but sent up sprouts from its base that were dug and planted by Roswell Humphrey, Chapin's brother-in-law. By 1840 the apple was attracting the attention of other growers, and in 1852 the American Pomological Society recommended it as an apple worthy of general cultivation.

Northern Spy is a vigorous tree with stout, arching branches that form a dense rounded tree that becomes pendulous with continued cropping. Standard trees can attain great size and in older orchards were often planted up to 50 ft. (15 m) apart. Although considered hardy, Spy will not perform well outside of Zone 4. Where temperatures fall below –22°F (–30°C) productivity will decline, and winter injury will weaken the tree. One of its more well-known faults is its reticence to come into bearing. On seedling roots it is common for Spy to take six to eight years to begin producing a crop. On new size-controlling rootstocks this wait time has been reduced substantially.

Northern Spy is a large fruit that is round to conic, flattened at the base and sometimes ribbed. The skin is thin, tender and smooth with a pale-yellow background that becomes overlain by pinkish red and red striping. Many love this apple eaten fresh as well as baked in pies. Northern Spy needs to be handled with care to maintain its quality in storage. If put into storage directly after picking it will keep in a cold cellar till late winter. Controlled storage allows it to keep until early the next summer.

Pie makers who use this icon of the apple world usually proclaim that they do so because those who know pies will gravitate in their direction. ■

Northwestern Greening

Hardiness Zone 3
Introduced 1872
Origin Waupaca County, Wisconsin
Primary uses Cooking, baking

The first tree of Northwestern Greening I ever saw was in the orchard of a century-old farm next to mine. The tree was mammoth with an immense trunk. The farmer was very proud of the tree, as he was of his entire farm, and rightly so on both accounts. I had just begun grafting historical apples, and he assured me no collection would be complete without this old-timer.

Northwestern Greening is generally credited to a cross of Golden Russet with Alexander. It was introduced by E.W. Daniels of Waupaca County, Wisconsin, in 1872. Although it bears scant resemblance to either, it certainly inherited great hardiness from Alexander. This tough tree can easily handle temperatures of –40°F (–40°C) or lower. It was the cooking apple of choice for much of the northern Midwest states and was popular in the colder sections of New England and Atlantic Canada in the 19th and early 20th centuries.

The grass-green fruits have noticeable raised white lenticels scattered across the surface. While usually oblong to slightly conic, the shape of the fruit can be somewhat asymmetrical or elliptical. It has a short stem. As it ripens, the fruit turns a pale yellow, often with a faint

Once grown in many northern orchards, this "Granny Smith of the north" is still very valuable as a cooking and keeping apple.

The tree is healthy and very hardy.

Northwestern Greening stays grass-green until fully ripe, when it will yellow a bit, often blushing softly.

reddish blush toward the sun. The even coloration and waxy complexion of Northwestern Greening would come a close second to Granny Smith for perfection. This fruit was never considered a dessert fruit, being somewhat too tart and firm. Its calling was as a baking and sauce apple, at which it excels. It will keep till midwinter or later under proper conditions. If not picked it will hang on the tree till spring.

This apple is used by the famous pie maker Mrs. Smith and is planted in small quantities in the northern states. Where Rhode Island Greening and similar green apples are too tender, Northwestern Greening should be considered a suitable replacement. ■

Novamac

Hardiness Zone 4
Introduced 1978
Origin Kentville, Nova Scotia
Primary uses Dessert, cooking

In Nova Scotia there is a valley renowned for its beauty and tranquility — a quilted, verdant stretch of green slung between two long ranges of hills. Here grow some of the finest apples in the world. The soils are rich, and the nearby Bay of Fundy moderates winter temperatures and keeps spring and fall frosts at bay. It is a perfect place for an orchard.

Until recently the most important apple grown in the Annapolis Valley was McIntosh. In warmer areas McIntosh forms a soft apple that does not keep well, but in the north, where nights are cool during ripening, the McIntosh grows to perfection. It seems only fitting that a child of McIntosh should have been born in the Annapolis Valley, where harvests of this special apple have kept food on fruit farmers' tables for so long.

At the heart of the valley, in the town of Kentville, where the Apple Blossom Festival decorates the town each spring, is the Kentville Research and Development Centre. Here research and breeding work on fruit trees have been conducted since the 19th century. When breeding for scab resistance began in the United States, the research center formed associations with universities that were conducting these programs. It received material from those programs, and from its efforts several promising selections were chosen. One was a cross between McIntosh, an apple noted for its superb flavor as well as an extreme susceptibility to apple scab,

Novamac captures the best of McIntosh without the disease susceptibility — a near-perfect apple.

The tree is broad, spreading and annually productive.

Even unripe Novamacs are tart and tasty. At peak they are sublime, but don't delay as the fruit will drop.

and a scab-resistant seedling called PRI 1018-3 (NJ 24 x PRI 47-147). The result of this cross was a perfect marriage of McIntosh's flavor and the numbered seedling's health. Novamac was released to the public by Dr. A.D. Crowe in 1978.

Novamac is a moderately vigorous spur-type tree that is naturally open with branches that grow from the trunk at a 45-degree angle, making it easy to prune. It begins bearing at a young age and is a consistent annual bearer of moderate to heavy crops. The foliage is glossy and does not suffer from apple scab, fireblight or cedar apple rust. We have found it resistant to codling moth as well. Two small faults are a tendency for terminals to occasionally split as they grow (bifurcation) and a tendency to drop fruit as it nears maturity. Novamac is hardy to –31°F (–35°C) and can survive –40°F (–40°C) with minimal injury.

The fruit is round to slightly conic, and its skin has a green background that becomes blushed red over nearly the entire surface, particularly where it receives direct sun. In southern New Brunswick it ripens around the beginning of October, but it can be picked several weeks earlier if you like your apples tart and later as well. If picked at the proper time, Novamac will keep in common storage till mid-winter and in controlled atmosphere storage till June. The flesh is greenish white to bright white, crisp and juicy. Like McIntosh the flesh is melting and the skin tender. Many would not be able to tell Novamac from McIntosh, although as it ripens Novamac becomes somewhat more vinous and has a richer aroma. It is superb as a dessert apple and makes a fine sweet cider.

Few apples have better credentials for organic production. Its hardiness, precocity, productivity, tree form, health, flavor, long season and storability are ideal for organic orchards. This is an apple that should be better known and one that will hopefully play an important part in orchards of the future. ■

Parkland

Hardiness Zone 2 (possibly 1b)
Introduced 1979
Origin Morden, Manitoba
Primary uses Dessert, cooking

Parkland is a 1979 introduction from the Morden Research and Development Centre that has proven hardy in the coldest of areas. It is grown in the Canadian prairies and has found a home in Alaska, where it is one of the most important cultivars for the growing number of commercial orchards there. Some report success in sheltered areas of Zone 1b. Parkland is a cross between the crab Rescue and Melba.

The tree is a compact grower and a moderately productive annual bearer with medium-sized fruits. The fruit has a yellow-green background and forms a soft red blush where exposed to sun. The flesh is white with a crisp texture and a pleasant but decidedly tart taste. It can keep for two months in cold storage.

Parkland now vies with Norland for the hearts and mouths of far northerners. Though such apples may not be as complex in flavor as their more southerly kin, they serve to add a richness to the lives of those living in the most challenging frontiers of apple growing. ■

Patten Greening

Hardiness Zone 3
Introduced 1885
Origin Charles City, Iowa
Primary uses Dessert, cooking

Planting apple seed is akin to buying a lottery ticket. Most often the fruit that appears after several years' growth is a bust. Occasionally you get one that makes you keep trying, but the big prize is elusive. Optimistic fools and relentless gamblers press on.

C.G. Patten of Charles City, Iowa, was such a gambler, but his bet paid off. He purchased a handful of seeds gathered from a Duchess of Oldenburg tree that a farmer named Daniel Eastman of Portage, Wisconsin, had grown. What eventually hung on the branches of Patten's new seedling was an apple of large stature. Though some are medium in size, most are 4 in. (10 cm) across and resemble lime-green pumpkins with very short stems nestled in a narrow, deep cavity. Around the cavity is a small starburst of russet. Where the sun is most intense, the skin blushes a dull bronzy-red. The lenticels are submerged; on the shaded side they are green, and on the sunny side they appear as tiny black dots.

But, as they say, size and, indeed, looks are not everything. This apple's flesh is moderately fine, yellow and crisp. Its plentiful juice is slightly tart and pleasant. Its skin disappears with a few chews. No doubt Patten, or more likely his wife, made pies with their crop, and they must have been pleased. Indeed, this old-timer is still a great multipurpose apple with the added benefit of being exceptionally hardy.

Patten must have felt as if he had won against the odds. He loved it enough to introduce his apple to the world, and over time his seedling, Patten Greening, found its way across America and up into the northern areas of the U.S. and Canada. It can still be found in many old orchards.

When fully ripe the fruit often has a soft blush on the sunny side.

The tree is large, open, somewhat pendulous and extremely hardy.

The tree is rather open with branching that is slightly more horizontal than vertical. Standing under an old Patten Greening tree is like being under a giant umbrella. It is moderately vigorous and is impervious to all but the most extreme cold. It has been used by breeders in the creation of other hardy cultivars, such as Goodland, thus extending the cultivation of quality apples northward and upward (if you live far above sea level). Patten Greening is also exceptionally resistant to apple scab and shows less insect damage than most.

The lure of the jackpot keeps people buying tickets week after week or, if you are an apple obsessive like Patten, keeps you on the search for the newest apple that will thrill the world. ■

Patten Greening has a classic oblate form.

Patterson

Hardiness Zone 2
Introduced 1960
Origin Saskatoon, Saskatchewan
Primary uses Dessert, cooking

With small- to sometimes medium-sized fruit, Patterson teeters on the edge of crabapple stature. The apples are generally found in clusters on a somewhat willowy tree. Toward summer's end, the skin of the apple morphs from green to greenish yellow, even looking a bit translucent as the fruit grows older. Occasionally a light blush appears on the sunny side. The stem is medium in length and very slender. When you bite into a Patterson, you are met with a snow-white, mildly tart flesh with a pleasant texture. Although such apples might be common in warmer climes, this apple can grow where the thunderous cracking of thick ice on a northern lake might be heard within earshot of the orchard — places where winters are dark and long.

The creation of such an apple tree was the goal of Dr. C.F. Patterson, a breeder working at the University of Saskatchewan. By 1946 this apple had been selected as having both excellent quality and superb hardiness. Though small, Patterson is ideal for dessert or cooking. When cut, the apple resists browning, making it appealing for salads. It also has juice aplenty for making an early sweet cider.

The tree is an annual producer of moderate to heavy crops. Thinning should result in somewhat larger fruit. The tree has wide crotches and will become rather sprawling over time. Heavy crops can break its branches. It might be wise to prune back the tips of outward-growing branches in summer while the tree is young. This will result in the creation of a more compact structure that will stand up to the weight. Its branching habit makes Patterson an excellent candidate for espalier. Growing against a south facing wall will speed ripening, an important asset in places where growing seasons are short.

This apple is most common in the prairie provinces of Canada and has also found a place in Alaskan orchards. While Patterson may not set the apple world on fire, it provides a bounty for those living where apple growing is a risky venture. ■

Paulared

Hardiness Zone 4
Introduced 1968
Origin Sparta Township, Michigan
Primary uses Dessert, cooking (sauce)

In 1960 Lewis Arends was walking along his block of McIntosh apples when he came across a wild seedling. One bite from one of its apples and he realized he had made an important find. By 1968 this new apple, presumed to be a seedling of McIntosh, was introduced to the world. Arends named it after his wife, Pauline.

Paulared is an early apple, ripening in mid to late August in most areas. Although not thought of as a keeping apple, in cold storage Paulared can be kept for up to two months.

It has a great tart-sweet balance, perhaps a bit heavier in the tart department. It is a fantastic early eating apple and is touted as an excellent cultivar for apple sauce, which will be pink in color and needs no sugar. It is a bit too soft for pies and crisps.

Paulared is often called an Early Mac, although there is another apple by this name, which was bred in Geneva, New York, as a cross between Yellow Transparent and McIntosh. The flavor of Paulared is certainly reminiscent of McIntosh. When picked at or just before perfect ripeness, it is crunchy with plenty of juice and has the acidity many of us crave in an apple. If left on the tree, many will drop and the remaining fruits will get soft.

The skin background is lime green with sun-drenched specimens turning 50 percent or more red, which is softened by a light bloom. When rubbed it becomes glistening red. The

Paulared's a great early apple that's worth growing despite a tendency to scab. No one's perfect.

The tree is rounded and annually productive.

lenticels are spread evenly across the fruit and are white and noticeable. The fruit is oblate to round with a medium-length stem.

Paulared, like its presumed parent, is subject to scab, and so prevention is necessary if you are going to grow these commercially. For the home grower the scab is not as consequential, especially for those making sauce or sweet cider. There are usually plenty of clean specimens for the table.

This is an important apple in the industry and has seen good staying power against many of the newer early apples. It should be around for many years to come. ■

Despite competition from many new early cultivars, Paulared has stood its ground.

Pewaukee

Hardiness Zone 4
Introduced ca. 1868
Origin Pewaukee, Wisconsin
Primary use Cooking

This apple originated as a cross between Duchess (some say a Duchess seedling) and Northern Spy (some say Blue Pearmain) made by George Peffer of Pewaukee, Wisconsin, sometime around 1842. In 1868 it won a $50 prize as the best new seedling by the Wisconsin State Horticultural Society and was named Pewaukee at the same meeting the next year. Although Pewaukee was grown in that area, it seems to have made more of an impact in places like Maine and New Brunswick, where it was valued for its hardiness and usefulness in cooking.

Pewaukee's tree is healthy and long lived. Its form is upright, becoming rounded with right angled branches that curve upward. Old specimens can still be found today. It produces annually and is relatively productive. Pewaukee flowers late and is a late apple, maturing in mid-October in New Brunswick. It is firm and will keep well into the winter.

The fruit is medium in size, near round, somewhat ribbed and with a nearly absent calyx cavity. Prominent protuberances at the stem end are common and an easy way to identify this old cultivar. The skin is greenish yellow, becoming blushed with orange-red and striped with carmine-red.

The flesh is yellow and coarse but very juicy and melts easily in the mouth. It makes a lovely tart sauce.

My nursery once had a contest to guess an apple we placed on a display table. The prize was $200 for the correct guess. I thought no one would recognize the specimen of Pewaukee I used. I underestimated the memories of an older couple who both guessed correctly. I was $400 poorer but a lot wiser about future guessing games. ■

Pomme Grise

Hardiness Zone 4
Introduced 18th century
Origin Quebec or France
Primary uses Dessert, cooking, cider

Pomme Grise is such a pretty little thing that it is a shame it has become rare. Its history is somewhat murky. Some say it was a seedling found in New France (Quebec) in the 18th century, others say it arrived on a ship from France. Whatever the truth, from there it spread out across much of the apple growing areas of North America and came to be known under a host of local names, including French Russet, Grise, Leather Apple of Turic, Leather Coat and Gray Apple. It took well to many sites, including the orchards of Thomas Jefferson's Monticello, where it was considered one of the top-quality apples. As size and packout became important for apple growers, Pomme Grise was slowly relegated to the discard bin of horticulture.

This is an apple with superb qualities. The flesh is crisp, finely textured, aromatic and juicy. The flavor has been described as rich, sweet and nutty with hints of cider, citrus, pear and vanilla. The small- to medium-sized fruit has a somewhat flattened globular form. It becomes quite handsome as the skin's yellow-green background transforms into a suede-like covering of tan russet with a gray overcast. On sun-drenched specimens a rosy blush can appear under the gray.

The tree is very hardy and can be grown through Zone 4 and possibly in the warmer sections of Zone 3. It is not an overly vigorous tree, which makes it an excellent backyard tree. It is an annual producer. This apple is making a bit of a comeback as a good cultivar for cider.

Size has been this apple's downfall, but skilled pruning can help produce at least medium-sized fruits. Besides, most apples, especially those destined for children, are far too large anyway. Let us support the size challenged and strike a blow for diversity in size. ■

Priscilla

Hardiness Zone 4
Introduced 1974
Origin West Lafayette, Indiana
Primary uses Dessert, cooking, cider

I was once approached by an elderly lady looking to purchase apples for jelly. She could no longer find the apples she usually used, and she specifically wanted an apple that would produce a pink jelly. I had been picking Priscillas that day and showed her their red skins and the wisps of red seeping into the white flesh. Her eyes gazed into mine as if to ask if they could possibly replace the apples she usually used. Suffice to say they did. Until she passed away, that lady came back for Priscillas every year. I think how simple and satisfying the joy she received from the smell and sight of apples bubbling in pots and jars of pink jelly.

Priscilla was born when pollen from a scab-resistant seedling was placed upon the pistil of a Starking Delicious flower and grew down to the waiting ovary. The uniting of these two apples created a seed that was gathered by breeders working in West Lafayette, Indiana. The year was 1961. In the spring of 1962, the seed grew into a tree, and by 1966 it was bearing fruit. What the breeders liked was the apple's crisp texture, sweetness, mild acidity and pleasing aromatics — these and the tree's complete resistance to apple scab. They introduced Priscilla to the world in 1974. It was one of the early introductions of the Purdue University, Rutgers University and the University of Illinois (PRI) breeding program for scab resistance.

Priscilla's fruit is oblate to round with a yellow background color that eventually shifts into a bright red that covers nearly the entire surface. Many remark that it is a comely apple. Though the size has limited any commercial uptake, Priscilla is a great apple for the home orchard. The fruit is ripe after the earliest apples, like Yellow Transparent, but before the

Priscilla is an early to midseason apple that does not get scab.

The tree is broad and spreading, and the fruit is well distributed throughout.

Priscilla makes a superb sauce or jelly, but it's also excellent for eating right off the tree.

later apples come in. Priscilla makes a terrific lunch-box treat, an excellent sauce and, of course, a lovely pink jelly. It also makes a balanced early-season sweet cider.

The tree is moderately spreading and has good vigor. It has withstood –40°F (–40°C) with aplomb. While it is often listed as a Zone 5 tree, it has no problems in Zone 4, and a risk-taker might try to grow it in the warmer sections of Zone 3. ■

Pristine

Hardiness Zone 4
Introduced 1994
Origin West Lafayette, Indiana
Primary uses Dessert, cooking

This superb early apple was chosen from a crowded row of seedlings planted in 1975 at Purdue University in West Lafayette, Indiana. All the seedlings had been screened in the greenhouse for susceptibility to apple scab and other diseases. Only the resistant seedlings made it to the field. However, they had to show their mettle as superb-quality apples, and few were chosen to be evaluated. Far fewer were given a name.

Originally named Co-Op 32 (a cross between Co-Op 10 and Camuzat), this apple stood out as a strikingly clean fruit. It is a medium to large apple with a classic oblate form and is flatter than most. The skin starts out bright green, gradually transitioning from soft to bright yellow, often with a smallish soft-orange blush where exposed to sun. There is no russet on the skin, nor are the lenticels prominent. The overall effect is one of glossy perfection. The name Pristine is well chosen.

Pristine's flesh is crisp and breaking yet melts in the mouth, yielding a perfect balance of tart and sweet. Those who have tasted it prefer it to Lodi, the early apple that is used as a touchstone for comparison with other early cultivars. Not only is the taste comparable and, in my opinion, superior, but also it will keep for four to six weeks in your refrigerator.

The tree is round topped and spreading, with branches that are well angled for strength and ease of pruning. The branches are limber and will droop under heavy loads, so cutting back the tips of branches (a process called heading back) to form stockier wood should be considered. The foliage seems impervious to apple scab and mildew with only a slight susceptibility to cedar apple rust and fireblight. Pristine has a tendency for biennial bearing. Thinning of fruit will create more annual production.

The list of new apples is long, but relatively little attention has been paid to early-season apples. Pristine is one of the more important new introductions for this season, and its beauty and superb disease resistance make it all the more exciting. ■

Pumpkin Sweet (Pound Sweet)

Hardiness Zone 4
Introduced ca. 1834
Origin Manchester, Connecticut
Primary uses Baking, cooking

In days of old every apple had its purpose. Unlike today, when virtually all apples sold in the supermarkets are dessert apples, many of the historical varieties were grown for a single reason. Pumpkin Sweet (alias Pound Sweet, Lyman's Large Yellow, Lyman's Pumpkin Sweet and others) was *the* baking apple. The tree originated in the orchard of S. Lyman of Manchester, Connecticut, and was first noted in 1834.

As its name suggests, Pumpkin Sweet is a very large apple when well grown. It is round to conic in shape, sometimes slightly elliptical and often with prominent ribs. It starts out grass-green, then changes to a clear yellow marbled with greenish yellow. Specimens grown in full sun may exhibit a subtle brownish-red blush but never a distinct red coloration. Some rays of russet can often be found in the stem cavity. The lenticels are white with a russet center. The stem is short and stout. The flavor of a freshly picked Pumpkin Sweet is unusual but agreeable, with low acidity and a distinct sweetness. It is quite crisp and juicy when fresh. The fruit will keep till midwinter if stored properly.

This apple shone for making baked apples which, alas, are not made as often as they once were. The core would be removed, filled with sugar and spices, then baked in the oven till soft. Vintage Virginia Apples tells of an orchard in Conneaut, Ohio, planted with Pumpkin Sweet that was used to make apple butter during the Civil War.

The tree is upright and very vigorous with hefty branches. It eventually forms a rounded, slightly open tree. It has proven hardy. Trees in good sites have survived nearly 100 years, even where temperatures have fallen to −40°F (−40°C). It tends to be biennial, but thinning can help alleviate this fault. The heavy fruit will drop in a severe wind, and if proper pH is not maintained, the fruit can develop watercore.

I would grow this apple for its name alone, but luckily it has enough attributes, not the least of which is hardiness, that make it a worthwhile addition to a northern orchard. ■

Red Astrachan

Hardiness Zone 3
Introduced ca. 1835 (in the United States)
Origin Russia
Primary uses Dessert, cooking

Often called simply Astrachan, this famous apple once graced apple orchards across the continent and was a favorite in northern areas because of its great hardiness. It was never a favorite of commercial growers, owing to its tender flesh and short shelf life. In short it is not a great shipper. For home orchards it was, and to a small extent still is, a wonderful apple.

Though most pomologists agree it was originally from Russia, the earliest record of this apple is a report of its introduction from Sweden into an English garden by a Mr. Atkinson in 1816. By 1835 it was being exhibited at the Massachusetts Horticultural Society's exhibition. Its use spread rapidly, and by the turn of the century Red Astrachan had found its way into the hearts of many in the New World.

The fruit is quite irregular in form, generally round to slightly oblate but often tending to conic and usually softly ribbed. The skin is tender with a yellow-green background becoming striped and then overlain with a red blush that can become quite dark and pervasive, especially when grown in full sun. A mature fruit will have a bluish bloom. A quick rub, and the fruit will shine. The flesh is white, but usually infused with red adjacent to the skin. On the tongue the fine, crisp flesh is juicy, slightly tart and aromatic. Cooks can pick this apple a bit early, when its acidity is higher, to make a great apple pie as early as late July or early August.

Red Astrachan forms an extremely vigorous upright tree that spreads with age. On seedling roots this tree can become 30 ft. (10 m) tall or more, so using a dwarfing

Red Astrachan is tart and tangy, but once ripe it will turn mealy fast.

The tree is among the most vigorous I know. Using size-controlling rootstocks would help with this cultivar.

rootstock will render this a more manageable tree. Although crops will be produced most years, it has a tendency toward biennial production. It is somewhat subject to scab and is very attractive to the apple maggot fly. You must use this apple quickly as it will turn rotten in a very short time.

You will never see rows of Red Astrachan flowing over the hills into the distance, but one tree in the kitchen garden will be a source of joy that will most likely outlive those who plant the tree. This means future caretakers can enjoy this old and venerable tree and partake of its delightful, delectable fruits. ■

The fruit is striped but can become a deep red when exposed to sun.

Red-Fleshed Crab (Hansen's Red Flesh)

Hardiness Zone 3
Introduced 1928
Origin Brookings, South Dakota
Primary use Cooking

The great South Dakota breeder Niels Hansen contributed in many ways to the diversity of fruit that could be grown in northern areas. Red-Fleshed Crab is another example of his unique approach to breeding. Using a cultivar of the native crabapple *Malus coronaria* called Elk and the red-leafed and red-fruited Russian species *Malus niedzwetzkyana* (a species that also helped create the popular pink flowering crabs), he created Red-Fleshed Crab, a small apple that is marvelous as a pickled or sugared crab or as the base of a lovely deep-red jelly.

This is one of the reddest apples you will likely come across. It is ovate in form with prominent knobs on the basin. The skin is a solid deep burgundy with no other color to mar its perfection. Underneath is a flesh that is nearly as red. Even the seeds are red. Though Hansen may not have realized it at the time, the high levels of red coloration are due to anthocyanins, pigments located in many plants that have exceedingly high levels of antioxidants, which are now seen as important for maintaining good health. Breeders are now busily at work crossing and introducing new red-fleshed apples, but Mr. Hansen beat them to it by nearly a century.

The tree is exceedingly hardy, handling temperatures of –40°F (–40°C) with aplomb. It forms a moderate-sized upright to slightly spreading tree with reliable production every year. The leaves and fruit are only mildly susceptible to apple scab, but apple maggot finds the fruit very attractive. With protection from this insect, the fruit is very handsome as well as nutritious. ■

Both the inside and outside of this crabapple turn deep red.

The tree is smallish and pendulous. Heavy crops can break branches with their weight.

Redfree

Hardiness Zone 4
Introduced 1981
Origin West Lafayette, Indiana
Primary use Dessert

This apple is the result of controlled crosses from the Purdue University, Rutgers University and the University of Illinois (PRI) breeding program, which was initiated to create scab-resistant cultivars. Redfree was the sixth selection to be released and remains to this day a superb introduction. Though perhaps not well known, it has a loyal following among those who do.

Redfree is an early apple; however, unlike many in this season, it will keep for several weeks in a refrigerator and maintain its crisp texture. The fruit is large and if it receives sunlight the waxy, glossy skin will color completely red. Its shape is round to conic with a knobby bottom somewhat reminiscent of Delicious. Redfree's flavor is mild and sweet with just a hint of tart. When bitten, the light-yellow flesh breaks apart with juice aplenty. The resulting sections, though seemingly coarse, melt in the mouth.

The tree is willowy, and training in the nursery is necessary to produce a straight trunk. Though hardy by most standards, Redfree is at its limit of hardiness in Zone 4. Winter injury will occur at temperatures of –31°F (–35°C), though it generally recovers quickly. When winter injury occurs, extra care must be taken to ensure that European canker does not move in. The tree, leaves and fruit are immune to apple scab, powdery mildew, fireblight and cedar apple rust.

As a testimonial to this fruit's flavor, we have sold it to several you-pick operations in Atlantic Canada. Years later many growers come back looking for more because their customers

single it out as their favorite. Additionally, because it is so early, the growers have been able to capitalize on the fall fervor for apples when most cultivars are far from ripe. For those in Zone 4b or warmer this is one of the finest of the disease-resistant early apples. ■

Rhode Island Greening

Hardiness Zone 4
Introduced Late 17th century
Origin Newport, Rhode Island
Primary uses Cooking, dessert

Sometime around 1650, among the tangled greenery in the yard of the Green's End Inn, a tavern in Newport, Rhode Island, an apple seedling appeared, perhaps planted from seeds brought from England. The tree flourished, and soon its large green fruits were being picked for pies and baked apples at the inn. It became apparent that an important new apple had appeared. Although initially slow to spread, within a century Rhode Island Greening had become the most important culinary apple in the colonies. It was said that the original tree nearly died from so many people taking scion wood.

Until the mid-20th century it retained its preeminent position, and even today Rhode Island Greening is grown in places like New York State for processing and is sought by those looking for a great pie apple. However, in a world of controlled storage facilities, culinary apples have taken a distant second place to dessert apples, and cultivars such as Rhode Island Greening are becoming less common. Several green apples have appeared since that are most likely seedlings of Rhode Island Greening — such as Bottle Greening — but they have not had the staying power of their parent and are now nearly obsolete.

This late-season apple is irregularly round to oblate. The skin is grass-green but often has rays of russet emanating from the stem cavity. Sun-drenched specimens will have a bronzy-red cast on the sunny side. The white lenticels are quite pronounced. Once seen, it is easy to pick out a Rhode Island Greening from the crowd. The flesh is yellowish, crisp, moderately tart and unlike any other in flavor. If left on the tree until the fruit turns slightly yellowish, the flavor will be superior. Even before the advent

This apple is perhaps the most famous pie apple in North America.

The tree is large and wide, and has shown good hardiness at our site in Zone 4.

Though only slightly prone to scab, it does attract sawfly and codling moth, so monitor closely.

of controlled storage, Rhode Island Greening could be kept till early spring in a cold cellar.

Although temperatures below –31°F (–35°C) can injure wood, it will survive in Zone 4 if grafted on hardy roots and given a protected site. Though it is often said that it will not grow north of Zone 5, my tree has endured –40°F (–40°C) during its nearly 40 years in the orchard and is turning into a mighty specimen. The tree is vigorous and spreading, eventually becoming slightly drooping. It is a triploid cultivar so it is useless as a pollinator and will require two other diploid cultivars for pollination. The foliage and fruit are slightly susceptible to apple scab and seem attractive to codling moths.

At the beginning of the 20th century this apple was considered among the finest of cooking apples. With care and good growing techniques this important apple can still hold its own with more modern cultivars. ■

Roxbury Russet

Hardiness Zone 4
Introduced Late 17th century
Origin Roxbury, Massachusetts
Primary uses Dessert, cooking, cider

This might be considered the great-granddaddy of American apples. It is considered by most to be the first named apple cultivar in America. There are records showing that trees or scions were taken from the farm of Ebenezer Davis of Roxbury, Massachusetts, where this apple originated, to Connecticut in 1649 — meaning the original tree must have first appeared in the 1620s or 1630s, when Europeans were first colonizing the area. This was undoubtedly grown from seed brought over by the first European settlers. It is a testament to its excellence that this apple is still grown today.

The Roxbury Russet is a medium to large oblate to oblate-conic fruit, sometimes elliptical. The stem is short to medium in length, thick and often shaded red on one side. The skin's greenish-yellow background is overlain with a patchy layer of yellow-brown russet, sometimes with a bronze blush that can turn to dull red. Inside is a green-yellow flesh that is coarse yet juicy and sprightly with overtones of spice. Although somewhat tart, there is a great deal of sugar, which the cider maker can take advantage of.

Roxbury Russet has done well in Zone 4. In colder areas be sure it is grafted on hardy roots. The tree is long lived and vigorous with strong limbs and a spreading form. This is an excellent pollinator for mid to late-season bloomers and is partially self-fruitful.

This apple, considered by many to be among the finest flavored of any, was originally planted as a cider apple. If stored in favorable conditions, the fruit will keep till the following summer — one of the reasons it was so popular in the days before refrigeration. It is also quite resistant to most diseases that plague apples, including apple scab, mildew and fireblight. Like Golden Russet, its complexity makes it a worthwhile addition to a cider mix. Though difficult to find nowadays, it is worth the effort of searching out. ■

One of the first named apples in North America, Roxbury Russet remains a first-class apple.

The tree grows as wide as it is tall. It is annual and productive.

The fruit is fairly resistant to scab and moderately resistant to insects.

Sandow

Hardiness Zone 4
Introduced 1935
Origin Ottawa, Ontario
Primary uses Dessert, cooking

This fabulous apple has been rejected by growers largely because of one flaw — it has an upright growing habit that makes it more difficult to prune. While certainly a justifiable reason for the commercial grower, the home gardener or adventuresome orchardist need not be so dismissive. This apple has many attributes that recommend it to our attention.

Sandow was released by Agriculture Canada. It is an open pollinated seedling of Northern Spy that was planted by D. Blair and B. Hunter (some sources credit W.T. Macoun) in 1898, first selected in 1915 and introduced in 1935. It has the all the attributes of its parent and more.

The fruit is medium sized and round to oblate, though some will tend toward conic and are noticeably ribbed. The green background gives way to stripes and blushes of red over a good portion of the surface, though shaded specimens will remain greener. Some will have a netting of russet. The stem is usually short and slender. The new growth of Sandow tends to be short and stout with tomentose (woolly) terminal buds.

The flesh is tart-sweet and more aromatic than Northern Spy, with a distinctive aroma of raspberry. At full ripeness the firm, coarse-textured flesh cracks when bit into, providing juice in spades. Sandow is superb as both an eating and culinary apple. Its flavor is a unique set of aromatics that places Sandow in a class by itself.

Though the tree does have a strong upright growth habit, early pruning and limb spreading can help create a more open form. The tree is hardier than its parent; it has proven perfectly hardy in Zone 4 and is well worth trying in the warmer sections of Zone 3. One of the

I think Sandow is one of the most intriguing flavors among apples, deserving far more attention than it has garnered.

Sandow is very resistant to scab, although maggots seem attracted to it.

The tree is upright and vase shaped. Early training will pay off.

best characteristics of this tree is its resistance to scab. When a local grower I know was purchasing an orchard that had been neglected and unsprayed for several years, he told me the Sandows stood out among the rows as clean and inviting. It seems somewhat resistant to codling moth, but unfortunately apple maggot finds it tasty. It is also said to be susceptible to fireblight, though I have not seen this in our orchard. Cold temperatures help keep this disease at bay, one of the few upsides to northern growing.

Though Sandow will never become an important commercial cultivar, let us hope it will still be grown in smaller orchards and backyards. Those who have bitten into a fresh ripe fruit or made a pie with it most often become instant converts. ■

Seek-No-Further (Westfield Seek-No-Further)

Hardiness Zone 4 (possibly 3)
Introduced ca. 1796
Origin Westfield, Massachusetts
Primary use Dessert

The world of apples is filled with wonderful names, such as Cow's Snout and Sheep's Nose, but perhaps one of the most optimistic is Seek-No-Further. How could one not include an apple with such a name? It conjures images of the discoverer walking along a hedgerow, reaching up to pluck an unprepossessing fruit and realizing on first bite that it was the apple they had longed to find. They would have to seek no further. Spencer Ambrose Beach, in his book *The Apples of New York* (1905), described its flavor as "very good to best" — a ringing endorsement, indeed.

This old dessert apple originated in or near Westfield, Massachusetts, in the Connecticut River valley. The first mention of it was in 1796. It was considered the apple of choice in that area and its cultivation spread throughout the northern United States, making its way to the Lake Ontario region and eventually to Atlantic Canada.

The tree is exceptionally hardy and can be grown with no injury in Zone 4 and possibly Zone 3. It forms a rounded, dense tree of average height. It bears annually and abundantly, and holds onto its crop till ripe with little dropping. Though not immune to apple scab, it is not overly susceptible and has good resistance to insects. References from Maine say it was "badly attacked by the apple maggot," but strangely we have found it one of the more

Seek-No-Further is a lovely dessert fruit. It's not as juicy as Honeycrisp, but its flavor is much more complex.

The tree is vigorous and rounded, producing an annual crop.

resistant. The codling moth, however, does seem to enjoy it.

The fruit has a tough greenish-yellow skin that becomes overlain with brick red and covered in a bluish bloom that gives the fruit a dull overcast. The lenticels are rather large and conspicuous. Its stem is characteristically thin and relatively long. The flesh is crisp and breaking with a rich aromatic flavor. It is not known as a culinary apple.

Although most gourmands today would not place Seek-No-Further among the best, many in the past considered it a great dessert apple, and some still think the same. At the very least it deserves an exalted place among the apples of our past. ■

Seek-No-Further is not immune but is quite resistant to scab and insects.

Silken

Hardiness Zone 4
Introduced 1999
Origin Summerland, British Columbia
Primary use Dessert

This is a suggestive name for an apple that seems to be living up to its billing.

Silken is an apple from the Pacific Agri-Food Research Centre in Summerland, British Columbia. It is a cross between Sunrise, an earlier release from the same institution, and Honeygold, a Golden Delicious cross from the University of Minnesota. It is interesting to note that in the early 19th century a Russian apple named Silken was introduced but apparently made little impact.

This is a striking apple with what has been described as porcelain skin. Its lustrous creamy skin is often softly blushed with rose but never striped. There is commonly some russeting in the stem bowl. The fruit is medium in size and somewhat oblong and ribbed. It is picked just before McIntosh, when the basin turns from green to white.

Taste tests that employed both professionals and us common folk rated Silken extremely high for flavor. It was considered crisper, sweeter and more aromatic than either McIntosh or Gala. Once its qualities are known to more people, it should prove to be a popular apple and will be easily recognized by its unique appearance.

This is a dessert apple for immediate consumption, though it will keep for several weeks in cold storage before losing quality. It is recommended that the trees be picked several times over its 10-day peak of ripeness.

The tree is spreading with an open spur-type habit, making it a good candidate for high density orchards. It is easy to train and very precocious, often fruiting in the second leaf. It sets heavily, so early thinning is highly advisable to produce good-sized quality fruit. It does not appear to be overly susceptible to diseases. It has proven hardy, though its usefulness in Zone 3 may be questionable.

We have yet to see how far this new star will rise. It certainly is a unique fruit and another example of how the consumer's acceptance of apples other than red has broadened the selection on the counters. This is a good trend. ■

SnowSweet

Hardiness Zone 3
Introduced 2006
Origin Minneapolis, Minnesota
Primary uses Dessert, cooking

SnowSweet has a depth of flavor that should make it better known.

One of the most prolific breeding programs for northern apples is at the University of Minnesota. The list of excellent new cultivars from this institution is staggering. Among the latest is SnowSweet, a cross between Sharon and Connell Red that was introduced in 2006.

The fruit is medium in size and round conic in shape. The skin has a green-yellow background that becomes almost completely shaded red and a noticeable bloom that gives it a soft look that turns glossy when rubbed off. Inside is a pure-white flesh that is largely sweet, balanced by a touch of tartness. Its flavor has been described by many as buttery and rich. Another characteristic is the apple's ability to resist browning when cut, making it ideal for salads or sliced apples.

The tree is moderately vigorous with an open, slightly pendulous habit. It is considered as hardy as Honeycrisp, so it should be suitable in Zone 3.

The buzz about this new apple is strong. For northerners it is yet another cultivar to add to the catalog of superior apples. If only Peter Gideon, the 19th-century Minnesotan breeder of Wealthy (the first apple to thrive in Minnesota's climate), could have seen where his beginning efforts at breeding have led. Few then could have imagined such fantastic advances in the quality of extremely hardy apples. ■

Trees are annual, productive and super hardy.

Spartan

Hardiness Zone 4
Introduced 1936
Origin Summerland, British Columbia
Primary uses Dessert, cooking

In British Columbia's Rocky Mountains, a series of narrow lakes fills the bottoms of steep north-south running valleys. The rain and fog of the Pacific coast have been spent on the western flanks of the Rockies by the time the westerly winds reach the Okanagan and Kootenay lake country. The land around the lakes receives scant rainfall and is more desert-like than forested, yet along the lakes you would never know. Where the mountains have eroded to form softly sloped plateaus are farms irrigated by the lakes and underlying aquifers. Many of the farms have orchards of apple, plum, peach and cherry. Increasingly, however, they are being uprooted and replaced by vineyards, for the climate is not only amenable to tree fruits. Here, the world's most desired grapes ripen in the protected valleys under near-cloudless skies, and you can find some of the finest wines grown in Canada.

Hugging the shore of Okanagan Lake is the Summerland Agricultural Research Station, now known as the Pacific Agri-Food Research Centre. For many years breeders at the station have been crossing apples, always in the hopes of improving the quality of the fruit and the possibilities for the farmers who grow them.

Spartan is extremely flavorful and a great apple for a kid's lunch box.

One of the first breeding lines involved crossing McIntosh, an apple known for its flavor and melting texture, and Newtown Pippin, an exceptionally flavored green apple. The finest result of these crosses was Spartan, which was released in 1936.

Spartan is a small- to medium-sized, round to somewhat conic apple with wavy knobs on the basin. The skin is a greenish yellow that becomes a deep, rich burgundy red with white lenticels scattered evenly over the surface. A bluish bloom enhances the darkness of the fruit. A quick rub, and the fruit glows. The flesh is almost pure white with the slightest of pink tones and has a fine-grained texture that melts in your mouth. It definitely has a McIntosh-like flavor but is more vinous — a distinctive flavor all its own.

The tree is thrifty and well formed, with sturdy branching at a good angle for pruning. It is a naturally small tree, a decided advantage in the orchard. Its leaves, like so many healthy trees, have a glossy sheen. Strangely enough, though McIntosh is very prone to apple scab, its offspring is among the least prone of any commercial apple. Lesions on both leaves and fruit are minimal. It has, however, inherited its parent's hardiness. Spartan can be grown in the cooler sections of Zone 4 and, with wind protection, perhaps in Zone 3b. It grows to perfection in Zones 4 to 6. It is late maturing and resistant to pre-harvest drop. Its firmness makes it a good keeper into midwinter.

Though only one seedling in a row of many, this tree must have been visited many times as its attributes drew the attention of those bearing clipboards, scales and pressure testers — the tools of apple matchmakers. More important tools, however, are the mouths and tongues — instruments that, in the end, determine the future of every new contender for horticultural nobility. ■

Spartan has great scab resistance, and the tree has proven exceptionally hardy.

Suncrisp

Hardiness Zone 4
Introduced 1994
Origin Cream Ridge, New Jersey
Primary uses Dessert, cooking

For northern growers the jury's still out on this newcomer. It is a cross between Golden Delicious and (Cortland x Cox's Orange), so it certainly has a lot of quality genes in its background. The cross was done at Rutgers University's Horticultural Research Farm and tested as NJ 55. Certainly in warmer areas this apple has already developed a rabid fan base.

Suncrisp is medium to large in size and round to conic in shape. Its lovely yellow skin is blushed orange-red — perhaps some of Cox's Orange's genes expressing themselves. In cooler and more humid climates it will be subject to russeting, most prevalently in the stem cavity. This is the characteristic that "plagues" its parent Golden Delicious at such sites. The crisp flesh has been described as fantastic, exciting, overwhelming and various other awestruck adjectives. The insides are greenish yellow and will keep their crackling texture for at least six months in controlled atmosphere storage. The skin will take handling better than Golden Delicious, which is good news for shippers.

The tree tends to be quite upright with the fruit buds evenly dispersed throughout the canopy. Suncrisp bears on both spurs and terminal wood. If not thinned aggressively it will tend to alternate heavy and light crops, but with thinning it is dependably annual. It is rated for Zone 4, but relatively little testing has been done in this zone and it is unlikely to be a candidate for any place colder.

Let us hope the hardiness of Cortland has had a hand in creating this new kid on the orchard block. Its quality certainly is not in question. ■

SweeTango

Hardiness Zone 3
Introduced 2009
Origin Minneapolis, Minnesota
Primary use Dessert

This apple comes from the University of Minnesota breeding program and was introduced in 2009. As has been done with numerous new cultivars, it is a "club apple," meaning that no one without a license can propagate this tree. The idea is that the royalties are managed and all go directly back to the breeder, in this case the university. This follows the direction of nearly all plant breeding today. Virtually every new release has royalties attached. There is a good argument to be made that breeders should be compensated for their work, which often takes years and sometimes does not end till long after the breeder is gone, though the system of plant royalties has created large corporations whose existence devours much of the money collected. But enough horticultural politics.

SweeTango is a marketer's dream. The round to conic apple has a yellow background that becomes nearly covered in a uniform blush of orangey-red. White lenticels are noticeable and evenly spaced over the surface. This is an apple derived from a cross of two new stars, Honeycrisp and Zestar!. It has the same explosive juiciness of its parents and captures that perfect balance of citrusy tart and brown-sugar sweet. The spicy aromatics here are subtle but present, making this an attractive apple for the modern consumer.

Growers should be aware that SweeTango is susceptible to scab and moderately susceptible to fireblight. Another issue for commercial growers is its thin skin. It must be handled with care when harvested. It is also prone to russet patches, which are caused by the rapid expansion of the fruit. When the skin stretches it can split. The scars that result heal into what we call russet. However, there is no problem with hardiness. SweeTango could be grown into Zone 3 without issue. It is a midseason apple, so its season is pretty much over by the time the winter apples appear. It offers a high-quality apple for this season and packs a crunch that holds the Guinness World Record for the loudest crunch of any apple. (I'm not kidding!) ■

Sweet Bough

Hardiness Zone 4
Introduced 1806
Origin Unknown, possibly Maryland
Primary uses Dessert, cooking, baking

This handsome apple has been known by at least 30 names in its long history, most having "sweet" as part of them. No one knows the exact origin of Sweet Bough, but by 1810 it was being offered in Baltimore, Maryland, and soon after in the Town of Flushing (now Queens), New York.

Although most of the crop is medium sized, some apples grow into impressive specimens. Sweet Bough is roundish to conic in form with a deep, narrow stem cavity. Its perimeter is gently furrowed, suggesting an overstuffed pillow. Within is a short thick stem that rarely rises above the top of the cavity. The basin is moderately deep with smooth edges that can be slightly furrowed. The skin is greenish yellow, sometimes faintly blushed with red on the sunny side. The lenticels are faintly perceptible soft green dots that appear submerged. The overall impression is of a smooth, evenly colored surface of sublime lemon lime.

Sweet Bough's shiny skin is rarely blemished by scab, and the tree is resistant to cedar apple rust. It is somewhat susceptible to fireblight.

Sweet Bough is an early-season apple that is acceptable for dessert. Picked too early, the fruit will seem bitter, but as the season progresses it becomes tender, juicy and quite sweet. Many describe it as honey sweet. Though one of the finer sweet apples for dessert, this apple has been used more as a cooking apple, though its lack of acidity can call for a dash of lemon juice. It might be best used for coring and baking.

It is worth noting that the flowers of Sweet Bough are somewhat larger than most and are intensely fragrant. The tree is upright when young, becoming spreading with age, and age it does well. I have seen an ancient tree in the Annapolis Valley of Nova Scotia whose mammoth trunk was split and looked barely alive, yet the tree was loaded with fruit and the ground beneath it covered in flawless yellow spheres. A romantic sight, but also a sign the tree does tend to drop its fruit when ripe. Sweet Bough is a precocious bearer of biennial crops. Use the apples quickly, as they will not keep for more than two weeks.

Though barely known today, this apple is an old-timer that has always found enough fans to keep it from falling into complete obscurity. With all the new early cultivars available today, Sweet Bough will have a hard time maintaining a place in the modern apple world. It was once the first choice for early apples, and it would be nice to see it survive into the future. ■

Sweet Sixteen

Hardiness Zone 3
Introduced 1978
Origin Minneapolis, Minnesota
Primary use Dessert

A bridesmaid of an apple, Sweet Sixteen has been given a back seat in the apple world. This notable University of Minnesota creation began as a cross made by Dr. William H. Alderman in 1936 between Minnesota 447 (Frostbite) and Northern Spy. In 1947 Alderman named his selection MN 1593. Testing continued on this apple till 1978, when it was renamed and released.

If grown well, Sweet Sixteen is a large round to conic apple that begins green, becoming striped and blushed red with raised lenticels that give it an unshaved feel. The light orange-yellow flesh has a crisp, juicy texture. Before fully mature it has hints of cherry. As it ripens there are notes of anise, bourbon, vanilla, nuts and spice. It is sweet but with enough tart to help balance the flavor. One can sense the influence of Frostbite in the flavor profile of this apple. Sweet Sixteen is clearly a worthwhile apple for our stable of hardy cultivars.

The tree has great vigor and is extremely hardy, growing well in Zone 3, and it also has remarkable disease resistance. It keeps for two months in common storage and can last for four months in controlled storage. This is one of great cultivars for the far north. Author Warren Manhart of Oregon, which has a warmer climate, includes it among the 50 cultivars he chose to include in his book *Apples for the 21st Century* (1995). Clearly this apple needs to be better known. ■

Tangowine

Hardiness Zone 3
Introduced ca. 1950
Origin Havelock, New Brunswick
Primary use Dessert

Who can resist a name like Tangowine? This apple was found in the late 1940s by Charlie Stultz of Havelock, New Brunswick. Stultz was a maker of sauerkraut, an orchardist, a self-taught horticulturist and a poet. He spent a goodly amount of time wandering the area around his home and found this tree amid a plethora of trees growing near the Havelock Lime Works. The high pH of the soil and good organics provided an ideal location for apple seedlings to thrive.

Tangowine is a medium-sized apple that is round to slightly conic in form. The skin matures to a solid deep burgundy red with a bluish bloom. It is among the darkest of apples. There is often a russet-colored suture line running from the stem cavity to the basin. The fruit has an unusually long and slender stem. The flesh is white but red streaks work their way in from the skin. It has a sprightly tang but perhaps is a bit sweeter than tart. Tangowine will keep till spring, even in common storage.

This is a late apple that is very resistant to apple scab. The tree has proven its hardiness, surviving without injury the terrible winter of 1981, when temperatures of 59°F (15°C) dropped within 24 hours to –4°F (–20°C). Tangowine has good vigor and forms an upright tree that widens with age.

This apple has remained fairly local to New Brunswick, but it has gradually spread from its origin and is well worth seeking out. It is the legacy of a keen mind that never stopped seeking new and marvelous things. ■

When fully ripe Tangowine is very dark red, often with a suture line running from stem to basin.

The tree is rounded and vigorous, with annual and moderate productivity.

The fruit has a rounded form with long stems.

Tolman Sweet

Hardiness Zone 4 (possibly 3)
Introduced 1822
Origin Dorchester, Massachusetts
Primary uses Dessert, cooking, cider

Many years ago I was approached by a man who was anxious that I visit his old orchard. He was convinced he had a Pumpkin Sweet. When I stood next to the tree, I found an apple that was green morphing to yellow, and along the surface ran a suture line from the stem cavity to the basin. Some of the fruits had a small patch of soft red blush, which does not occur on true Pumpkin Sweet. There was no doubt in my mind that this was a Tolman Sweet. Well, the man could not be convinced that his tree was not a Pumpkin Sweet, and he went to his final resting place assured he was correct.

This old apple, originally called Tolman Sweeting, is believed to have been found in Dorchester, Massachusetts, in the late 1700s. Scions were sent in 1796 to the Putnam Nursery in Marietta, Ohio, where it was grown and introduced in 1822. The apple sleuths believe it to be a cross between Sweet Greening and Old Russet, though this is guesswork.

The fruit is medium in size at best and is round oblate to somewhat round conic, occasionally elliptical. Its skin becomes entirely soft lemon yellow with scattered russet lenticels. Many specimens display a suture line, which is a key identification feature. There will often be rays of russet emerging from the stem cavity, giving the rather tough skin a rough feel. The flesh is fine textured, hard when newly picked and quite sweet. It is only moderately juicy. It was esteemed for, as the name suggests, its overwhelming sweetness and aromatics, reminiscent of pear and perhaps pumpkin.

This was much to the taste of apple lovers in the 19th century, and Tolman Sweet was grown in many orchards. The tree is extremely hardy and long lived with a wide pendulous form. It bears good crops annually and is somewhat resistant to the fungal scourges of apples. It is still in demand today, though more for its sugar content, which can run over 14 percent. This makes it an ideal apple for a hard cider blend, as the high sugars will translate into naturally sourced alcohol and give balance to the bitters and sours of the other apples in the mix.

The old-timers use it in baking and drying. Sauces will not need any sugar added with this apple. Tolman Sweet is definitely a top contender for the sweetest apple of all. ■

Viking

Hardiness Zone 3
Introduced 1969
Origin Rutgers, New Jersey
Primary uses Dessert, cooking

One of the fallouts from breeding programs are seedlings that do not pass the selection process. Most are destroyed, but some of those numbered seedlings, usually in places where the testing has taken place, find devotees and/or growers who feel their advantages make them viable domestic or commercial apples and should not be discarded as a result of one disadvantage. In this case it was the Purdue University, Rutgers University and the University of Illinois (PRI) breeding program to develop scab-resistant cultivars.

Turns out Viking, the very first apple to be selected in the program, did suffer from some scab and was not released, but growers who were testing this new seedling sang its praises as an early dessert apple of uncommon quality. It was also very hardy, and growers in colder areas found it did well. Viking was unofficially introduced at the Wisconsin Agricultural Experiment Station in 1969.

I have had feedback concerning Viking from growers in Newfoundland, a land that challenges the best of fruit growers. They have been very happy with this apple. It can be ripened in Newfoundland's short season and has the quality of flesh the public craves. Although it does not seem to have been adopted by the larger industry, this apple can be found in many northern orchards.

Viking is an apple whose fine-grained cream-white flesh has aromatics that make fresh eating a very satisfying experience and whose

Viking is a superb early eating apple with high sugars.

Tree is moderate in size and upright. It is annually productive.

fragrances are kept even after cooking. Aromas of rose and lychee have been noted by pommeliers. It qualifies as an excellent cooker, retains its shape and has a good balance of tart and sweet in either a pie or crisp. Its fragrance makes it outstanding for sauce. This is also a great apple for early sweet cider, which could be blended with bitter and astringent apples for hard cider.

The tree is upright but with pendulous branches once it is bearing crops. The form is open, allowing for lots of light, which makes most of the fruit a deep maroon, often with a bloom as it reaches peak ripeness.

Viking ripens after Yellow Transparent but before the midseason apples.

Although it can get scab, Viking is not what most would call a scabby fruit. Most apples are quite clean — far better than the average.

This apple came to us originally from a true northern orchard pioneer named Eddy Dugas. Dugas lived in the far northern corner of New Brunswick, where winter temperatures reached −40°F (−40°C) or lower nearly every year — a challenging place to grow apples. He spent a good portion of his life testing apple cultivars, rootstocks and growing methods. He compiled his accumulated knowledge into a publication called *La culture de la pomme dans le nord* (1992). The book was written in French, and it is a pity it has never been translated into English. It is people like Dugas who broaden our knowledge and stimulate us to keep searching for better. ■

Wealthy

Hardiness Zone 3
Introduced ca. 1860
Origin Minnesota
Primary uses Dessert, cooking

Peter Gideon was a pioneer fruit grower from Excelsior, Minnesota. He was the first to test apple cultivars in that northern region and record his results. He requested seed from Albert Emerson of Bangor, Maine, who had been working with seedlings and grafts of recently imported material from Russia. Out of his batch of seedlings Gideon chose one that was uncommonly hardy, very healthy and, for the time, superior in flavor to any other apple grown in northern Minnesota. He named it Wealthy.

Wealthy quickly became the most important apple in the northern Midwest and was soon being planted in large numbers in the Northeast, particularly in northern New England and Atlantic Canada. Even today many old trees, some of which were planted in the 1890s and early 1900s, are still producing fruit. Wealthy is definitely a tough tree.

The tree is less vigorous than many, becoming spreading and slightly drooping with age. Although not immune to scab, Wealthy is one of the more resistant of the older cultivars. It comes into maturity just before the late-season apples.

The fruit of this apple is round to slightly conic in form. The skin is streaked red, becoming heavily blushed when in full sun. The flesh is white, sometimes tinged with red, and is sprightly and aromatic, becoming sweeter as the fruit matures. It is considered a very good dessert apple, and many who know it consider it among their favorites.

This old-timer has withstood the test of time and is still worth planting. I imagine its name did not hurt its rapid dissemination. I am not sure whether any of the growers grew wealthy because of this apple. In my experience apple growers, or any farmers for that matter, are rarely wealthy. Farming is an act of love as much as economics. ■

Wickson Crab

Hardiness Zone 3
Introduced 1944
Origin Humboldt County, California
Primary uses Cooking, dessert, cider

Albert Etter was one of those individuals who was dedicated to the goal of creating new and different fruits but, because he operated outside the breeding establishment, was rarely taken seriously. I think also of Elmer Swenson of Wisconsin, whose work on hardy grapes was dismissed by the so-called experts but remains today among the most astonishing body of work on the subject.

Etter lived in Humboldt County, California, where he concentrated on producing apples with red flesh, using various crabapple species, in particular *Malus niedzwetzkyana,* a red-fleshed crab that had been introduced from Russia and was to become central to the introduction of both pink- and red-flowered ornamental crabapples. Wickson Crab is not, however, one of Etter's red-fleshed crabapple crosses. It is believed he crossed Newtown Pippin and Spitzenburg Crab to create Wickson. Etter introduced it in 1944 and named it after one of the few pomologists who took his work seriously, Edward J. Wickson.

This rather small and hardy tree produces large (for crabs) 1.5 to 2 in. (3 to 5 cm) fruits that are yellow with orange-red striping as they mature. The stem is rather long, and the fruit nearly round. Perhaps the most astonishing aspect of this crab is not that it is tart, as are most crabs, but that it has a 25 percent sugar content.

Although it can be eaten off the tree, this crab is highly esteemed as a preserving crab and has rocketed up the list of desired cider apples, as it offers acid, sugar and flavor. Of course the downside is that small apples require large numbers to give you any amount of volume. Nonetheless, Wickson is among a small elite of preferred cider apples today.

It gives me great joy to speak of great things accomplished by an individual with unlimited energy and imagination. ■

William's Pride

Hardiness Zone 4
Introduced 1987
Origin West Lafayette, Indiana
Primary uses Dessert, cooking

This is a new early-season cultivar from the Purdue University, Rutgers University and the University of Illinois (PRI) breeding program for breeding disease-resistant apples. Formerly known as Co-Op 23, the apple was selected by Dr. Edwin Williams of Purdue University from field trials because of its outstanding characteristics.

William's Pride ripens one week after Lodi, an early-season Yellow Transparent cross that is used as a measure of earliness. It is a medium- to large-sized fruit, oblate to round in form and softly ribbed. Its somewhat waxy skin changes from greenish yellow to a deep, rich red with tones of purple. This is further intensified by a heavy bloom that gives the apple a blackish-red look. The flesh is crisp and juicy with an excellent tart-sweet balance, one of the highest-quality apples in its season. It ripens over a long period and is best picked as it ripens.

The tree is vigorous, becoming spreading with age. It seems to be very healthy with few or no disease problems. It has shown no apple scab infection or cedar apple rust in all the years it has been tested and is rated as having good resistance to fireblight and mildew. Some bitter pit has been noted in certain years. So far it has proven hardy in Zone 4 and is worth trying in Zone 3. William's Pride bears a moderate to heavy crop annually, and its fruit hangs well even after ripening.

This is another exciting addition to what is a growing list of healthy apples. We can only hope this list will continue to expand, making apple growing an easier and more inviting activity for all. William's Pride is definitely a prime candidate for the organic grower's orchard. ■

Wolf River

Hardiness Zone 3
Introduced ca. 1870
Origin Wolf River, Fremont, Wisconsin
Primary uses Baking, cooking

If this apple had been small perhaps no one would have paid attention to it, but it is far from small. Wolf River is one of the largest apples grown. There are several accounts of how this apple came to be, but one seems to ring most true.

A lumberman from Quebec named William Springer decided to move his family to Wisconsin. While traveling along the shore of Lake Erie, he purchased a bushel of large red apples that he later came to believe must have been Alexanders. He told his family to save the seeds so they could be planted on their new homestead. They settled along the Wolf River in 1862 and planted the seeds. This particular seedling grew on the bank of the river, and when it began to produce apples, the cultivar was named Wolf River.

The quality of Wolf River has been disparaged by many. It cannot be denied that the flesh is rather coarse and cannot compete as a dessert apple with today's gourmet choices, but it has been primarily used as a baking apple. It keeps its shape when cored and filled with sugar, nuts and such. It has also been said that a pie could be made from a single apple, and that is not far from the truth.

The fruit, which does resemble Alexander, is very large, often up to 8 in. (20 cm) across. It is round-oblate and often asymmetrical. The skin starts as greenish yellow, but as it matures it becomes nearly covered in brick red, mottled with darker striping. The stem is very short and deeply recessed in the apple's narrow stem cavity. A burst of russet always surrounds the cavity, and a few rays of russet can sometimes reach to the base. Prominent russet lenticels are distributed

Wolf River is known for its size. Not a great eating apple, it is typically used for pies and baked apples.

The tree is more horizontal than vertical, with vigorous, thick stems.

evenly across the surface. The calyx is set deeply into a narrow basin, and the base of the apple is usually flat but can have a fleshy protuberances.

The tree is very vigorous and upright in youth but becomes a spreading tree of moderate size and eventually somewhat pendulous. It is not precocious, taking five or six years to begin producing. It blooms early in the spring, and although most years it produces a crop, the crops are heavier in alternate years. Wolf River is quite scab resistant and is reportedly resistant to cedar apple rust. It is also said that its seedlings will often resemble the mother tree.

Wolf River was a mainstay in historical northern orchards, partially for its size but mostly because it can survive the worst winters. Some may scoff at this old-timer, but in the end it has outlasted the naysayers and still has a legion of fans. ■

Wolf River is quite resistant to disease.

Yellow Bellflower

Hardiness Zone 4
Introduced Mid-18th century
Origin Crosswicks, New Jersey
Primary uses Dessert, cooking

A Yellow Bellflower by any other name might be Belle Flavoise, Belle Fleur Jaune, Belle-Flower, Bellflower, Gelber Bellefleur, Lincoln Pippin, Metzger's Calville or Bishop's Pippin. So many names indicate the widespread popularity of this apple. Found originally around 1742 on a farm in Crosswicks, New Jersey, its fame and grafted children eventually spread completely across the continent and northward to the lower latitudes of Canada.

My first introduction was to the Bishop's Pippin. It seems the Yellow Bellflower first fruited in the Annapolis Valley of Nova Scotia in the orchards of Bishop Charles Inglis. The bishop started as a poor, unconnected Irish-born boy who immigrated to America and through determination and hard work became an important man in New York City. The American Revolution forced this ardent royalist back to England, where he was made the Anglican bishop of Nova Scotia. He sailed to Nova Scotia, became the first headmaster of King's College and built a residence in the Annapolis Valley. He formed the first agricultural society there and promoted the growing of apples, helping to create one of the most important fruit-growing areas of eastern North America.

The bishop likely would have learned about Yellow Bellflower while in New York City, which is across the river from New Jersey, where this apple was first planted in the decades before his exile. Having arrived in Nova Scotia via England, it seems unlikely he brought trees or scion wood with him (though it is possible he brought such to England when he fled New York). It seems somewhat more likely that he sent for scions or trees from those he knew who grew this apple, but we may never know.

In any case, Yellow Bellflower became known in Atlantic Canada as Bishop's Pippin, pippin referring to a seedling. One wonders if he sought to claim it as his own seedling. The bishop was obviously a man of great ambition. Yet perhaps his fervor for the apple was such that no one paid attention to the name by which he knew it. Whatever the truth, it became Bishop's Pippin by association, and the name stuck. The local naming of apples is a common thread among nearly every old cultivar. It makes for fascinating history and confusing horticulture.

The fruit of this apple can vary tremendously in form. It is a medium to large fruit, usually conic in form, with a notably narrow and knobby basin — though some specimens can resemble a fat plum with a more blunt and wavy basin. They are often asymmetrical and ribbed. The color is usually a lemon yellow but can be blushed soft red where the sun has touched the fruit. The skin has numerous russet-colored lenticels scattered evenly across the surface. The color can intensify in storage to a deep golden yellow. Its skin is not built for rough handling, and care must be taken to not let newly harvested fruit sit in warm temperatures or the apples will not keep well. Unfortunately

this apple is susceptible to scab and, without attention to prevention, the scab lesions will affect storability.

Inside each apple is a tingling, tart flesh that is crisp, firm and juicy with a notable vinous aroma. When first picked Yellow Bellflower is ideal as a culinary apple since it is a bit tart for dessert. This tartness will balance out by developing sugars as the apple sits in storage. By midwinter it makes a luscious eating apple.

Yellow Bellflower grows into an imposing tree. It becomes very tall on seedling roots, though its branches droop gracefully. It is hardy in Zone 5 and on hardy roots can be grown into Zone 4b, though severe winters that test an apple's toughness might occasionally damage the wood.

I get my scion wood from a tree that looks out on the Bay of Fundy, a place where winds throw wicked tantrums but where winter temperatures are moderated by the ocean. It is perhaps among the most northerly trees of this cultivar in existence. It was on this unlikely spot that one of New Brunswick's oldest nurseries stood in the 19th century. The tree I visit was grafted from a yet older tree that was perhaps 150 years old when it finally succumbed to old age, grown no doubt by that pioneering nurseryman. It makes me wonder what name he knew this apple by. ■

Yellow Transparent

Hardiness Zone 2
Introduced 1870 (in the United States)
Origin Russia
Primary use Cooking

If you decide to grow apples in the north, the name Yellow Transparent will eventually arise. This is a cultivar with faults, the most significant being its biennial nature, yet it is still spoken of with reverence by those who remember the older orchards. It is the apple that won the north when few apples could overcome the harsh conditions. It is a survivor.

Hailing from Russia, where it had been grown for many years, Yellow Transparent, also known as Transparent or August Apple, was introduced in North America in 1870. Its superior hardiness quickly became evident, and it spread rapidly both north and west until it was found in nearly all northern orchards. Yellow Transparent is very early, usually ripening in August. The fruit is medium sized, though larger specimens can be had if grown under good conditions. It is round to conic with notable knobs on the basin. The skin is grass-green, becoming translucent yellow when ripe, with only wisps of russet in the stem cavity. The original Russian name is *Beliy Naliv*, which translates to "Ripe White." The overall effect is a smooth evenly colored fruit. The flesh changes from crisp and tart to soft and sweet in a matter of days.

As is apparent, this is not a dessert apple, though it is certainly pleasant to eat when at its peak. Transparent has always been considered a sauce apple *par excellence*. When it softens, the sugars increase, making the addition of sugar an option rather than a necessity. With

Pick the fruit when the green is morphing to yellow. Yellow Transparent makes the finest sauce there is.

The trees are smallish but sturdy. Few cultivars are hardier.

Yellow Transparent is relatively resistant to scab.

tartness to balance the sweet, you have an exceptional product.

Yellow Transparent is a small tree with a flowing branching pattern. Once observed, this cultivar can easily be picked out in an orchard. Indeed, in surviving orchards from the early 1900s, you will nearly always find at least one, for this is among the toughest of the tough, surviving temperatures of –49°F (–45°C) or colder. Identifying characteristics of the tree include its stubby fruiting spurs and plump buds. Though certainly not immune to scab, Transparent is not overly susceptible, and because it is generally used in cooking, a few lesions are immaterial.

Yellow Transparent will never be a commercial cultivar since it is biennial and breaks down far too quickly, but this pioneer apple will never let the home grower down. It deserves to be kept in the pantheon of great northern apples. ■

Zestar!

Hardiness Zone 3
Introduced 1999
Origin Minneapolis, Minnesota
Primary use Dessert

In common with many new introductions, Zestar! is a patented apple, meaning you cannot propagate this tree without a license from the University of Minnesota, where it was produced and introduced in 1999. Its trademark name comes with the "!," which I find a bit silly (but that's just me).

Zestar! is a medium- to large-sized round apple with a shiny green skin that shades to deep red where exposed to sun. Its lenticels are quite prominent. In keeping with the popularity of apples with a crisp, juicy texture, this apple provides both those qualities in spades, along with some aromatics that give it a hint of brown sugar. It is a lively fruit that satisfies, but it is not complex. One of Zestar!'s best attributes is its ability to keep well, even at room temperature. In fact, its flavor actually intensifies when kept past picking. In the refrigerator it will keep for two months.

The tree is moderately vigorous in its early life, but it slows down as it comes into production. Most commercial growers graft this apple onto size-controlling rootstocks to force early production. Zestar! is susceptible to scab and moderately susceptible to fireblight.

This has had good uptake in many areas, especially in the far north where it is one of the hardiest early apples on the market, doing well even in Zone 3 sites. It remains to be seen if it will become a major player. ■

Zestar! has a color that makes the apple industry swoon.

The tree is productive, annual and Minnesota hardy.

The fruit is medium to large in size and keeps well. ▸

Cider Apples

The following cultivars are grown primarily for the production of hard (alcoholic) cider. Most would not be considered good dessert apples or, for that matter, cooking apples. A good hard cider is a balance of sweet, bitter and sour. Thus good cider apples will have one or more of these qualities.

Cider has been made from time immemorial wherever apples could grow. In Europe the production of cider was centered in the counties of southern England, Brittany in France and the Basque Country of northwestern Spain. Over time trees were discovered that yielded exceptional ciders. Grafters created orchards of these select trees and enabled the cideries to create reliably high-quality ciders. Many of the better-known European cultivars have been imported into North America and often make up a good percentage of the trees planted. Though the qualities of these European apples are well known in the areas where they originated, their hardiness limits are still under evaluation, as is their suitability for cider making in colder climates.

In the early years of colonial North America, nearly all cider was made using seedlings, most descendants of the first seeds carried by boat to the continent. Soon orchards of seedlings and wild apples appeared wherever they would survive. As in Europe, apples were found that made exceptional quality cider. These were propagated, and such cultivars were planted in cidery orchards. With the near-complete demise of the cider market, brought on by the ascendency of beer and wine, most cider cultivars perished from neglect, were cut down or died a slow death in the shade of higher trees.

Today individuals have appeared, as if from nowhere, to take up the cause of saving the last surviving trees of such apples. By reintroducing them into the orchards of cider makers, these individuals are aiding a new generation of apple fermenters to both create and feed a resurging interest in cider and cider making. Many are dedicated to the craft of cider making, and these special older apples are critical to attaining the quality they strive for.

There are also new cultivars found and named by an esoteric group of cider makers that haunt the hedgerows and meadows in search of wildings that have the qualities needed for great cider — whether that be bitterness, sweetness, tang or any combination thereof.

I have mentioned quality cider apples throughout this chapter, particularly ones that also double as good dessert or cooking apples. These apples include Ashmead's Kernel (page 148), Chestnut Crab (page 161), Golden Russet (pages 186–187), Liberty (pages 199–200), Pomme Grise (page 224), Roxbury Russet (pages 236–237), Tolman Sweet (page 252) and Wickson Crab (page 256).

The cultivars included in this section are those known to be desirable for cider and that have been shown to have good hardiness in Zone 4 and some in Zone 3. Though it includes 11 cultivars that are great for cider, this list is still limited, as there are many new and often obscure apples that are just beginning to arrive on the scene.

A basket of Liberty, a quality hard cider apple. ▸

Banane Amère

Hardiness Zone 3
Introduced ca. late 1990s
Origin Quebec

Banane Amère was found in Quebec in the late 1990s by Claude Jolicoeur, author of *The New Cider Maker's Handbook* (2013), perhaps the most complete and valuable cider-making guide ever published. The original tree was found growing on his property. It was a multistemmed tree so tall that he could only harvest it when the apples fell. This is probably because it was an older tree competing for light with its neighbors. Still, a size-controlling rootstock would make sense for this vigorous cultivar.

Banane Amère has a very tannic bittersweet taste, has good body, is not very juicy and has a slight banana aroma. The fruit is round to oblate, greenish yellow to yellow in color and sometimes lightly blushed. It is slightly susceptible to scab. This is an extremely hardy mid to late-season apple.

Jolicoeur recommends only using 10 percent of this apple in a cider mix to avoid making too tannic a product. ■

Bilodeau

Hardiness Zone 3
Introduced ca. 1990s
Origin Lotbinière, Quebec

Bilodeau is a large crabapple that was sent to Claude Jolicoeur from a Monsieur Claude Bilodeau of Lotbinière, Quebec, in 1986. It has a high sugar content but is considered a bittersharp, meaning it has both bitterness and high acidity. It is an early-season crabapple with yellow skin and a red blush. It has an oblate form and grows to approximately 1.75 in. (4.4 cm) across. Some years it can show a netting of russet on the surface. The tree is extremely hardy with good scab resistance. Bilodeau is an annual bearer, though off years are not as productive. Thinning could help even out production.

It is possible this is the same cultivar as the Trail Crab. The form and slender stem are very similar, but Bilodeau blushes redder than Trail. ■

Brown's Apple

Hardiness Zone 4
Introduced Early 20th century
Origin South Devon, England

If you were to wander the cider orchards of England, Brown's Apple would no doubt be part of many orchards. This cultivar originated in South Devon, England, in the early 1900s. It is unsuitable for eating or cooking, but some consider it capable of producing a varietal cider, meaning this single cultivar can produce a fine, balanced cider. However, most use Brown's Apple to add acidity to the mix. By itself this apple will produce a thin cider, so it is recommended you add another apple to the blend that will give better body to the product. Brown's Apple is high in sugar, a quality necessary for producing alcohol. It has some bitterness and is highly regarded for its fragrance.

The tree is very vigorous and becomes spreading with age. It is extremely productive but more so in alternate years, as it does have a biennial tendency. The fruit is medium in size and oblate with a firm flesh. The skin becomes nearly all red with large soft pink lenticels with dark centers. Brown's Apple is quite late, which makes it unlikely to properly ripen in colder areas. It looks like Zone 4 will be its northern limit. ■

Bulmer's Norman

Hardiness Zone 4
Introduced Early 20th century
Origin Normandy, France

I have grown no other cider apple with the vigor of Bulmer's Norman. I am always impressed by the thick new growth that sets this cultivar apart from all others. This may in part be due to the fact that it is a triploid. So far it has been completely hardy in my Zone 4 orchard and may even be adaptable to the warmer sections of Zone 3.

As the name suggests, the origin of this apple is in Normandy, France. It was brought to England by cider maker H.P. Bulmer & Co. of Hereford in the early 1900s as an unnamed seedling and was planted in the company's orchards. It did not appear to be a named cultivar in France and was eventually named Bulmer's Norman.

The fruit is medium to large and conical or oblate conical in form with a somewhat waxy yellow to green skin, occasionally blushed orange-red. It has noticeable irregular-shaped lenticels. The stem cavity is wide and deep, with a short to medium stem. The basin can vary from shallow to deep and is often puckered and woolly looking, a characteristic called pubescence. This is a midseason apple and it is susceptible to scab.

Bulmer's Norman is considered a medium-bittersweet apple with hard and bitter tannins and some notes of mango and pineapple. It produces a goodly amount of juice and ferments rapidly. It is much in demand by growers as a blending apple because if used alone it is deficient in sharpness. ■

Dabinett

Hardiness Zone 4
Introduced Late 19th century
Origin Somerset, England

The first records of this apple come from the late 1800s in the Martock-Kingsbury area of Somerset, England. Dabinett is an apple unsuitable for any other use but cider, but as a cider apple it excels. Many consider this a good varietal.

The apple matures quite late. It is a small- to medium-sized fruit with waxy yellow to greenish-yellow skin, usually becoming blushed, flecked and striped with dark red. Its flesh is creamy to slightly green in color, somewhat crisp with sweetness and high astringency. The stem cavity is small and narrow, and the basin is shallow, usually smooth but sometimes puckered. It is very disease resistant.

The tree is small, precocious and heavy bearing. It eventually becomes spreading and is thinner wooded than many apples. Time will tell how far north it will grow. Dabinett will probably not do well north of Zone 4. ■

Douce de Charlevoix

Hardiness Zone 3
Introduced ca. late 1980s
Origin Baie-Sainte-Paul, Quebec

Douce de Charlevoix was discovered by Claude Jolicoeur in Baie-Sainte-Paul, Quebec, in the late 1980s. He took scions from a tree he believed was a grafted tree, but it turned out the original graft had died, and the scions were from the rootstock. The conical apple is medium to large in size with a yellow-green background that is partially striped and slightly blushed red. Jolicoeur claims the apple is quite resistant to scab. It is medium-bittersweet with good flavor.

An early apple, Douce de Charlevoix can be paired with similar early cultivars, such as Bulmer's Norman or Bilodeau. The apple produces a French-style cider with good mouthfeel. The tree is very hardy and bears annually. ■

Geneva Crab

Hardiness Zone 3
Introduced ca. 1930
Origin Ottawa, Ontario

This crabapple was developed at the Central Experimental Farm in Ottawa, Ontario, by Isabella Preston, a horticulturist who created a series of pink-flowered crabapples that came to be known as the Rosyblooms. They were also called the Lake Series, as they were named after Canadian lakes. Preston pollinated flowers of the pink-blossomed *Malus niedzwetzkyana* with the pollen of the much hardier *Malus baccata*. Unfortunately the purple fruit that resulted from her pollinated flowers turned out to be too tempting for passersby, and all but four were taken. From those four fruits came the seedlings that formed the Rosybloom series and gave us Geneva Crab.

Having escaped being eaten, this crabapple was nearly lost again. The original tree died, but it turned out that an entomologist, William Haliburton, who worked at the Experimental Farm, had taken cuttings and grafted trees for himself and a few others who liked the tree. His daughter later brought the farm some scion wood from her father's tree, and thus it was reintroduced to the world.

Geneva has medium- to large-sized (for a crab) round oblate fruit with reddish-purple skin and flesh that is deep pink from the skin inward, turning white around the core. Although on the tart side, it makes a good eating apple for those who eschew the sweet ones.

This crabapple has found a place in the hard cider world as a fabulous source of red color and a great source of tartness. It can be blended with sweet and bitter apples to create a quality cider.

Geneva Crab is a most lovely fruit. The flesh is nearly as red as the glossy skin.

It has a puckered basin.

The tree is partially self-fruitful but will benefit from the pollination of others. It will not pollinate other apples. The tree is quite striking with deep-pink blossoms and mahogany-toned foliage.

Geneva Crab does well in Zone 4 and should be hardy in the warmer sections of Zone 3. It ripens in mid to late September in the north, earlier farther south. The fruit hangs well in the tree with few drops. Good news for organic growers as well — Geneva is very resistant to scab. ■

Harrison

Hardiness Zone 4
Introduced Late 18th century
Origin Essex County, New Jersey

Harrison is an apple of legend. It originated as a seedling bought by a Mr. Harrison of Essex County, New Jersey, from a Mr. Osborne of South Orange, New Jersey, in the early 1700s. Several whips (I assume grafted whips) were planted out in his orchard, and the apple was eventually recognized as a premium cider apple. Harrison named it after himself. One wonders if Mr. Osborne felt slighted.

Its fame eventually spread as far as the American prairie states. As the production of hard cider dwindled, so did the growing of Harrison. For a time it was considered extinct, but in the fall of 1976, Paul Gidez, a Vermont orchardist and amateur apple sleuth, discovered an ancient tree close to where there had been a cider mill near the city of Newark, New Jersey. When he first began searching, he stopped by a bagel shop and asked if there had been any cider mills in the area. He was directed to the Nettie Ochs Cider Mill in Livingston and there he found an immense Harrison. He took scions from the tree the next spring. Shortly after that tree was cut down to make room for a garden. Because he had found it so easily, Gidez assumed there must be a lot of Harrison trees in the area, but further searches did not result in any being found until Tom Burford of Virginia found a second tree in 1989 on an estate near Paramus. This tree died the following year. A third tree was found near Newark in 2015 by Thomas Vilardi. It is chilling to think how close this tree came to becoming extinct.

Newark Cider was considered a close second to Champagne in the late 1700s. It was a blend of four apples: Harrison, Poveshon, Campfield and Granny Winkle. Apparently George Washington was an ardent fan. The Poveshon apple was also considered extinct, but a chance reading of its description by a New Yorker named Wesley Stokes changed that. He had apples in his old orchard that seemed to fit the description, and thus it was rediscovered. Now all four of the Newark cider apples are being propagated again and Newark Cider is once again available for cider connoisseurs.

The Harrison tree is vigorous, long lived and quite hardy. It is also extremely productive, helping to make up for its diminutive size.

This apple is one that might go unnoticed in the wild. It is small to medium in size and is yellow with blackish spots. The form is ovate to roundish oblong. Harrison's flesh is yellow, dry, firm and rather tough but with a rich subacid quality. It is considered among the highest in sugars. Harrison produces a highly colored cider with great body. We can thank intrepid fruit explorers for bringing this wonderful apple back in the fold again. ■

Kingston Black

Hardiness Zone 4
Introduced ca. 1826
Origin Somersetshire, England

Kingston Black is a medium-sized apple named for its deep red coloration, though it is not actually black in appearance. Kingston Black was grown extensively in western England until recently and was touted as a premier cider apple. It is a bittersharp with some tannins that makes an excellent varietal cider. There are, however, detractors who say that without blending the cider will be thin, and that the apple is best used to provide color.

The tree is strongly biennial and is only a fair producer, though production does seem somewhat dependent on where it is grown. On seedling roots it will take several years to come into bearing, so most growers use size-controlling rootstocks to shorten the wait time.

Although once popular, it has fallen out of favor with many growers because of its susceptibility to apple scab and canker. This may be more of a problem in colder areas, as winter injury invites canker. With all the problems that come with this apple it is hard to recommend it, but its stature as a first-class cider apple makes it impossible to ignore. ■

Muscadet de Dieppe

Hardiness Zone 4
Introduced Early 18th century
Origin Near Dieppe, France

Muscadet de Dieppe is an old French cultivar that is still used today to make high-quality cider. This is an early-ripening apple. It has some faults, such as dropping early in dry summers and low productivity, yet it is making inroads in North American cideries for its bittersweet quality and aromatic nature. The apple is relatively resistant to scab and fireblight.

Muscadet de Dieppe is greenish yellow with a noticeable orange blush and some striping. The tree is adaptable to various soil types and has shown hardiness in Zone 4, making it a promising choice for northern cideries. ■

Yarlington Mill

Hardiness Zone 4
Introduced Early 20th century
Origin Yarlington, England

Yarlington Mill is the most impressive of the European cider apples we have seen in terms of vigor. The sleek and deeply dark new growth seems to glow with health, though its leaves and fruits are somewhat susceptible to scab, mildew and fireblight.

An early 20th-century English apple, Yarlington Mill has a near-perfect blend of bitter and sweet, making it an ideal varietal cider. As interest in cider continues to expand exponentially, there will be ciders that, much like varietal wines and scotches, will be sought. Yarlington will undoubtedly be one.

The fruit has a round-conic form. The yellow-green background becomes lightly striped and turns pinkish red where sun hits. There is usually a bit of russeting in the stem cavity. The tree has proven quite hardy, but it needs a long season to ripen its fruit. Growers should thin crops to prevent biennial bearing.

The fruit is conic in form with green skin that blushes red. Sun-drenched specimens turn nearly all red.

The original tree was found in 1898 growing out of a wall at a water-powered mill in the village of Yarlington, Somerset, by a Mr. Bartlett. It was then promoted by Harry Masters, a well-known grower in the region.

If Yarlington Mill lives up to its reputation, northern cideries may have a quality source of juice for the future. Some have already run with this apple, as there are now many in the Northeast that are fermenting crops of Yarlington. ■

Yarlington Mill is one of the most sought-after cider apples and can create a superb varietal cider.

Appendix

◂ Bottle Greening apples ripening on the tree.

Canada's Extreme Minimum Temperature Zones

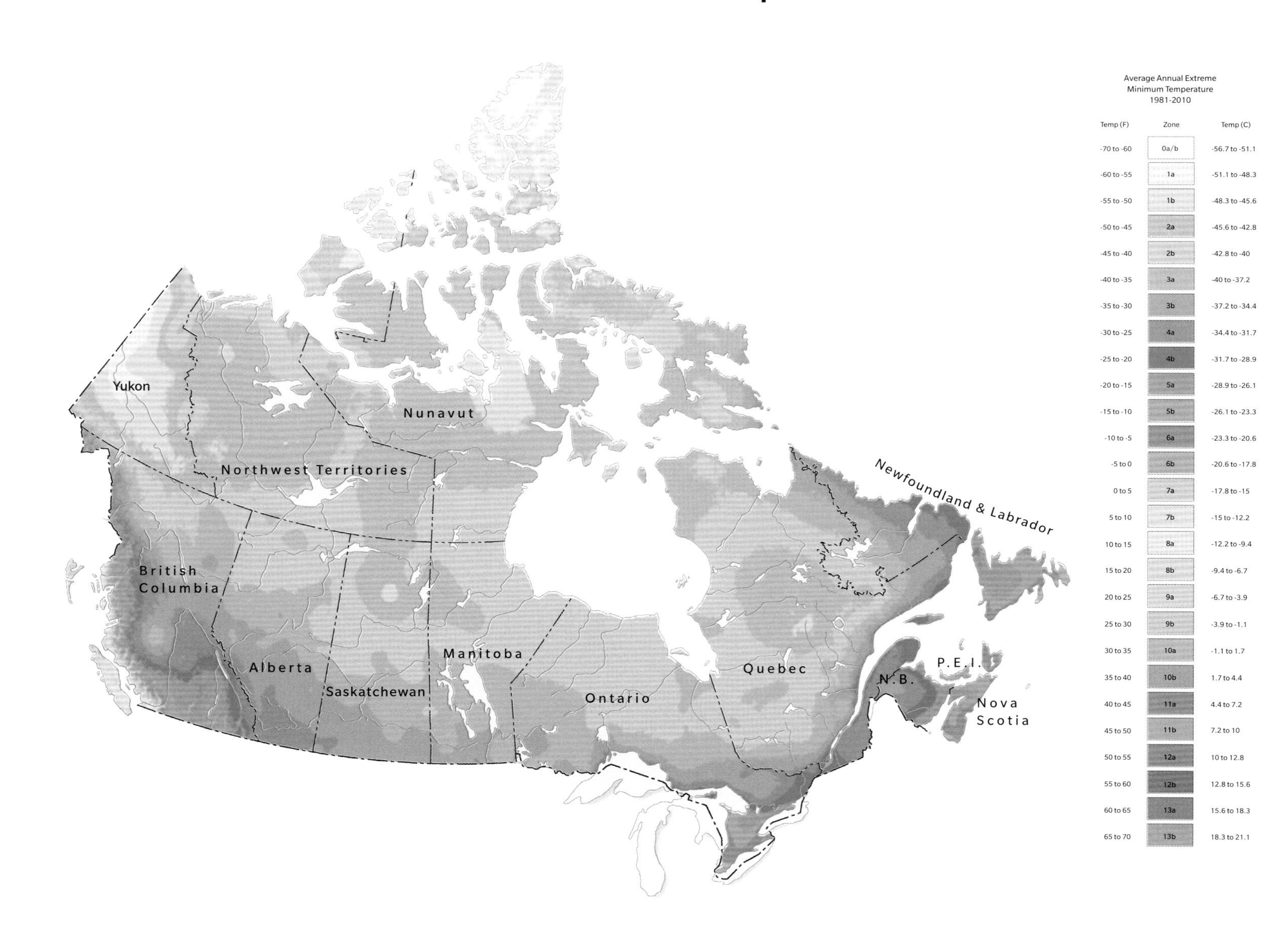

USDA Plant Hardiness Zone Map

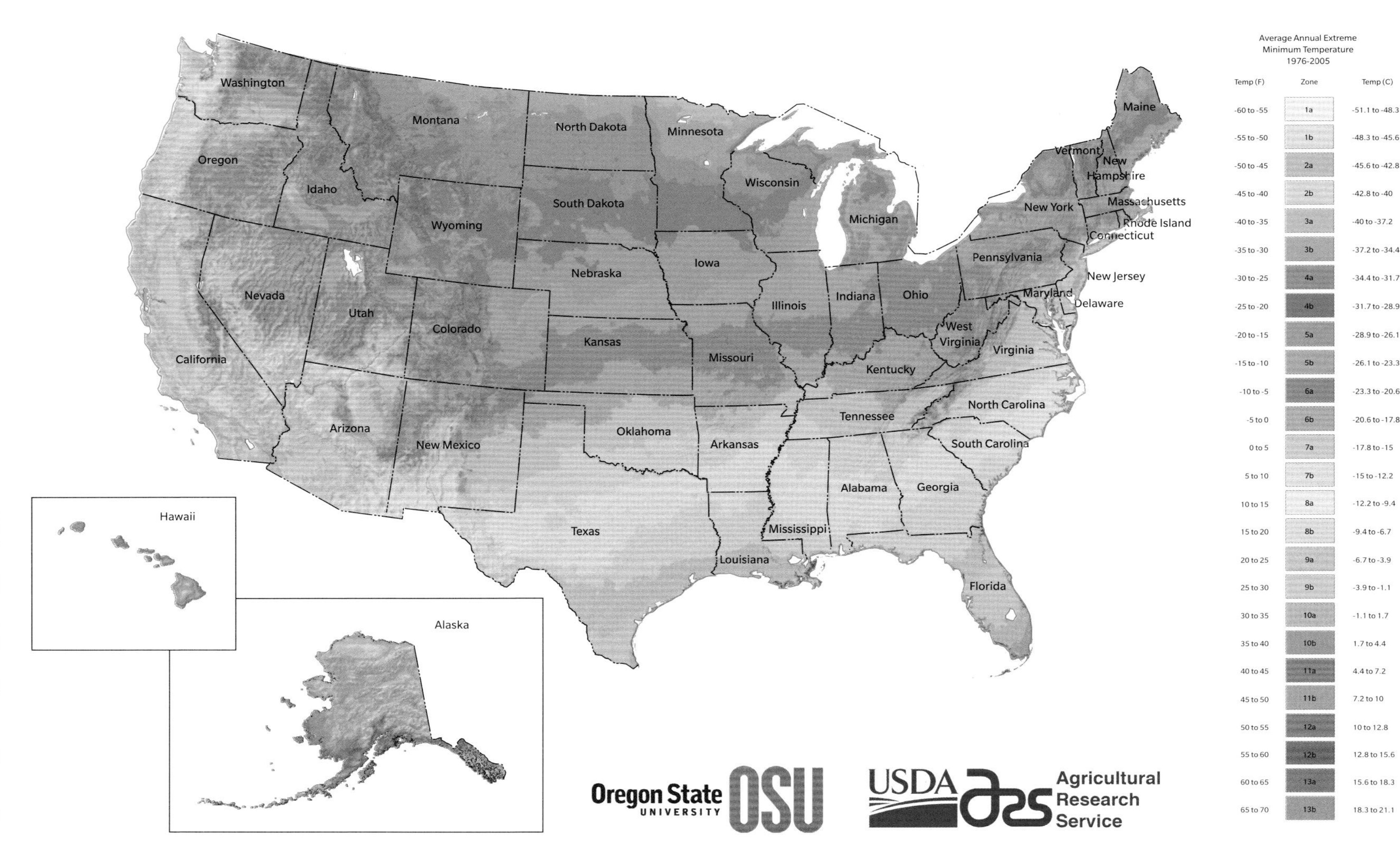

List of Hardy Cultivars

THE CULTIVARS	DESSERT	COOKING/ BAKING	STORAGE	BLOOM TIME	RIPENING TIME	ANNUAL/ BIENNIAL	FRUIT SIZE	DISEASE RESISTANCE	HARDINESS ZONE
Alexander	Fair	Good	1 Month	Mid	Midseason	Annual	Very Large	Fair	3
Ambrosia	Excellent	Fair	1 Month	Mid To Late	Midseason	Annual	Medium	Good	4
Antonovka	Good	Excellent	2 Months	Early To Mid	Mid To Late Season	Annual	Large	Very Good	2
Ashmead's Kernel	Excellent	Good	6 Months	Late	Late Season	Annual	Small To Medium	Fair	4
Bailey Sweet	Good	Very Good	1 Month	Early To Mid	Mid To Late Season	Annual	Medium	Fair	4
Ben Davis	Fair	Good	8 Months	Mid To Late	Late Season	Annual	Medium	Good	3
Bethel	Very Good	Very Good	6 Months	Mid To Late	Late Season	Annual	Medium To Large	Good	3
Black Oxford	Good	Good	8 Months	Mid To Late	Late Season	Annual	Medium To Large	Good	3
Blue Pearmain	Very Good	Fair	3 Months	Mid	Late Season	Biennial	Medium To Large	Good	4
Bonkers (NY 73334-35)	Very Good	Unknown	Unknown	Early To Mid	Late Season	Annual	Medium To Large	Excellent	4
Bottle Greening	Good	Excellent	6 Months	Mid	Late Season	Annual	Medium To Large	Very Good	3
Bramley	Good	Excellent	8 Months	Mid	Mid To Late Season	Annual	Large To Very Large	Very Good	4
Carroll	Excellent	Fair	3 Months	Mid	Early To Midseason	Annual	Medium To Large	Good	3
Chestnut Crab	Excellent	Good	2 Months	Early To Mid	Midseason	Annual	Small To Medium	Very Good	2
Cortland	Excellent	Excellent	8 Months	Mid To Late	Late Season	Annual	Large	Fair To Poor	3
Cosmic Crisp	Excellent	Very Good	8 Months	Mid To Late	Mid To Late Season	Annual	Medium To Large	Excellent	3
Cox's Orange Pippin	Excellent	Very Good	4 Months	Mid To Late	Late Season	Annual	Medium	Good	4
Crimson Crisp	Good To Excellent	Good	6 Months	Mid To Late	Midseason	Annual	Medium	Very Good	4
Delicious (Red Delicious)	Very Good	Fair To Goof	8 Months	Late	Late Season	Annual	Medium To Large	Good	4
Dolgo	Very Good	Excellent	1 Month	Early To Mid	Midseason	Annual	Small To Medium	Very Good	2
Duchess (Duchess of Oldenburg)	Very Good	Very Good	1 Month	Early	Early To Midseason	Annual	Medium	Good	2
Dudley (Dudley Winter)	Very Good	Very Good	3 Months	Early To Mid	Mid To Late Season	Annual	Medium	Good	3
Empire	Very Good	Good	5 Months	Early To Mid	Late Season	Annual	Medium	Good	4

THE CULTIVARS	DESSERT	COOKING/ BAKING	STORAGE	BLOOM TIME	RIPENING TIME	ANNUAL/ BIENNIAL	FRUIT SIZE	DISEASE RESISTANCE	HARDINESS ZONE
Enterprise	Excellent	Very Good	6 Months	Mid To Late	Late Season	Annual	Medium To Large	Excellent	4
Fameuse (Snow Apple)	Excellent	Good	2 Months	Mid	Mid To Late Season	Annual	Small To Medium	Fair To Poor	3
Fireside	Excellent	Excellent	7 Months	Mid	Late Season	Annual	Large	Good	3
Freedom	Very Good	Very Good	4 Months	Mid To Late	Late Season	Annual	Large	Excellent	4
Frostbite	Excellent	Good	4 Months	Mid To Late	Late Season	Annual	Small To Medium	Good	3
Gala	Excellent	Very Good	8 Months	Mid To Late	Late Season	Annual	Medium	Very Good	4
Ginger Gold	Excellent	Good	2 Months	Mid	Early Season	Annual	Medium	Poor	4
Golden Delicious	Excellent	Very Good	8 Months	Mid To Late	Late Season	Annual	Medium To Large	Good	4
Golden Russet	Excellent	Fair	6 Months	Early To Mid	Very Late Season	Annual	Small To Medium	Good	4
Goodland	Very Good	Very Good	5 Months	Early	Midseason	Annual	Medium To Large	Good	3
Granite Beauty	Very Good	Very Good	4 Months	Mid To Late	Late Season	Annual	Medium	Good	4
Greensleeves	Excellent	Very Good	4 Months	Early To Mid	Mid To Late Season	Annual	Medium	Excellent	4
Haralson	Very Good	Good	6 Months	Early To Mid	Late Season	Annual	Medium	Fair	3
Honeycrisp	Excellent	Very Good	8 Months	Mid	Mid To Late Season	Annual	Large	Good	3
Honeygold	Very Good	Very Good	5 Months	Mid To Late	Late Season	Annual	Medium	Fair	3
Hudson's Golden Gem	Excellent	Good	8 Months	Mid	Late Season	Annual	Large	Very Good	4
Hyslop Crab	Good	Excellent	1 Month	Early To Mid	Early To Midseason	Annual	Small To Medium	Very Good	3
Keepsake	Very Good	Good	7 Months	Mid To Late	Late Season	Annual	Small To Medium	Good	3
Liberty	Excellent	Very Good	3 Months	Early To Mid	Late Season	Annual	Large	Excellent	4 Possibly 3
Lobo	Very Good	Very Good	4 Months	Mid	Midseason	Annual	Medium To Large	Fair	4
Lodi	Good	Very Good	1 Month	Early	Early	Annual	Medium To Large	Good	3
Macoun	Excellent	Good	3 Months	Mid To Late	Midseason	Annual	Medium	Fair	4
Mann	Good	Very Good	6 Months	Mid To Late	Late Season	Annual	Medium To Large	Excellent	3
McIntosh	Excellent	Very Good	6 Months	Mid To Late	Late Season	Annual	Medium To Large	Poor	3

THE CULTIVARS	DESSERT	COOKING/ BAKING	STORAGE	BLOOM TIME	RIPENING TIME	ANNUAL/ BIENNIAL	FRUIT SIZE	DISEASE RESISTANCE	HARDINESS ZONE
Milwaukee	Fair	Excellent	5 Months	Mid	Mid To Late Season	Annual	Medium To Large	Good	3
New Brunswicker	Very Good	Excellent	2 Months	Early	Early To Midseason	Annual	Medium	Good	3
Norland	Very Good	Very Good	1 Month	Early To Mid	Midseason	Annual	Medium	Good	2
Northern Spy	Very Good	Excellent	8 Months	Late	Late Season	Annual	Medium To Large	Good	4
Northwestern Greening	Fair	Very Good	6 Months	Mid To Late	Late Season	Annual	Medium To Large	Very Good	3
Novamac	Excellent	Very Good	4 Months	Mid	Mid To Late Season	Annual	Medium To Large	Excellent	4
Parkland	Very Good	Good	2 Months	Early To Mid	Early To Midseason	Annual	Medium	Good	2 Possibly 1b
Patten Greening	Very Good	Very Good	4 Months	Early	Mid To Late Season	Annual	Large To Very Large	Very Good	3
Patterson	Very Good	Very Good	1.5 Months	Early	Early	Annual	Small To Medium	Very Good	2
Paulared	Excellent	Very Good	3 Months	Mid	Early To Midseason	Annual	Medium To Large	Fair	4
Pewaukee	Fair	Good	4 Months	Mid	Late Season	Annual	Medium	Very Good	4
Pomme Grise	Excellent	Good	5 Months	Early To Mid	Late Season	Annual	Small To Medium	Very Good	4
Priscilla	Very Good	Very Good	2 Months	Mid	Midseason	Annual	Medium	Excellent	4
Pristine	Excellent	Good	2 Months	Mid To Late	Early Season	Annual	Medium	Excellent	4
Pumpkin Sweet (Pound Sweet)	Fair	Very Good	5 Months	Mid	Late Season	Annual	Large To Very Large	Very Good	4
Red Astrachan	Very Good	Very Good	1 Month	Early	Early Season	Annual	Large To Very Large	Fair	3
Red-Fleshed Crab (Hansen's Red Flesh)	Fair	Very Good	2 Months	Early	Midseason	Annual	Small	Very Good	3
Redfree	Excellent	Good	2 Months	Early To Mid	Early Season	Annual	Large To Very Large	Excellent	4
Rhode Island Greening	Very Good	Excellent	6 Months	Mid	Late Season	Annual	Large	Good	4
Roxbury Russet	Very Good	Very Good	6 Months	Mid To Late	Late Season	Annual/ Biennial	Medium To Large	Very Good	4
Sandow	Excellent	Excellent	6 Months	Mid To Late	Late Season	Annual	Medium	Very Good	4
Seek-No-Further (Westfield Seek-No-Further)	Very Good	Fair	6 Months	Mid To Late	Late Season	Annual	Medium	Very Good	4
Silken	Very Good	Good	1 Month	Early To Mid	Early To Midseason	Annual	Medium To Large	Good	4
SnowSweet	Excellent	Very Good	2 Months	Mid	Late Season	Annual	Medium	Good	3

THE CULTIVARS	DESSERT	COOKING/ BAKING	STORAGE	BLOOM TIME	RIPENING TIME	ANNUAL/ BIENNIAL	FRUIT SIZE	DISEASE RESISTANCE	HARDINESS ZONE
Spartan	Excellent	Very Good	5 Months	Mid	Late Season	Annual	Small To Medium	Very Good	4
Suncrisp	Very Good	Good	6 Months	Mid	Mid To Late Season	Annual	Medium To Large	Good	4
SweeTango	Very Good	Good	2 Months	Mid To Late	Early To Midseason	Annual	Medium	Good	3
Sweet Bough	Very Good	Good	2 Weeks	Early	Early	Biennial	Medium To Large	Very Good	4
Sweet Sixteen	Excellent	Good	2 Months	Mid	Early To Midseason	Annual	Medium	Good	3
Tangowine	Very Good	Good	5 Months	Mid	Late Season	Annual	Medium	Excellent	3
Tolman Sweet	Very Good	Very Good	2 Months	Early To Mid	Late Season	Annual	Medium	Very Good	4 Possibly 3
Viking	Excellent	Excellent	1 Month	Mid	Early To Midseason	Annual	Large To Very Large	Very Good	3
Wealthy	Very Good	Very Good	4 Months	Early To Mid	Mid To Late Season	Annual	Medium	Good	3
Wickson Crab	Very Good	Very Good	2 Weeks	Early To Mid	Late Season	Annual	Small	Very Good	3
William's Pride	Excellent	Very Good	1 Month	Early To Mid	Early To Midseason	Annual	Medium	Excellent	4
Wolf River	Fair	Very Good	2 Months	Early	Midseason	Annual/ Biennial	Very Large	Very Good	3
Yellow Bellflower	Excellent	Excellent	5 Months	Mid	Late Season	Annual	Medium	Fair	4
Yellow Transparent	Good	Excellent	2 Weeks	Mid	Early Season	Biennial	Medium	Very Good	2
Zestar!	Excellent	Fair	1 Month	Early	Early Season	Annual	Medium To Large	Good	3

Bibliography

No person can know it all. The following books and websites have been invaluable in filling in the blanks in my knowledge. I am deeply in debt to the wisdom of the authors of the books listed here.

Beach, S.A. et al. *The Apples of New York: Report of the New York Agricultural Experiment Station for the Year 1903*. 2 vols. Albany, NY: J.B. Lyon Company, 1905.

Bunker, John. *Apples and the Art of Detection*. Self-published, 2019.

Bussey, Daniel J. *The Illustrated History of Apples in the United States and Canada*. Edited by Kent Whealy. 7 vols. Mount Horeb, WI: Jak Kaw Press, 2016. (This is the most complete list of North American apples ever assembled. An invaluable resource.)

Dugas, Eddy R. *La culture de la pomme dans le nord*. Saint-Leonard, NB: Éditions Dugas, 1992.

Hall-Beyer, Bart, and Jean Richard. *Ecological Fruit Production in the North*. Trois-Rivieres, QC: printed by the author, 1983.

Jacobsen, Rowan. *Apples of Uncommon Character*. New York: Bloomsbury, 2014.

Manhart, Warren. *Apples for the 21st Century*. Portland, OR: North American Tree, 1995.

Page, Stephen, and Joseph Smillie. *The Orchard Almanac*. Rockport, ME: Spraysaver Publications, 1986.

Phillips, Michael. *The Apple Grower: A Guide for the Organic Orchardist*. White River Junction, VT: Chelsea Green Publishing Company, 2005.

Stilphen, George Albert. *The Apples of Maine*. Bolsters Mills/Harrison, ME: Stilphen's Crooked River Farm, 1993.

Taylor, H.V. *The Apples of England*. London: Crosby Lockwood & Son, Ltd., 1936.

Upshall, W.H., ed. *History of Fruit Growing and Handling in United States of America & Canada, 1860–1972*. University Park, PA: American Pomological Society, 1976.

Yepsen, Roger. *Apples*. New York: W.W. Norton & Company, 1994.

Zielinski, Quentin Bliss. *Modern Systematic Pomology*. Dubuque, IA: W.C. Brown Company, 1955.

A Few Useful Websites

Agriculture and Agri-Food Canada: agr.gc.ca

Cornell University Fruit Resources site: fruit.cornell.edu

Iowa State University Department of Horticulture: hort.iastate.edu

Orange Pippin: orangepippin.com

Oregon State University Extension: extension.oregonstate.edu

Nova Scotia Fruit Growers' Association: nsfga.com/about-us

Pennsylvania State University Extension: extension.psu.edu

Pomiferous Apple Database (an excellent resource with information on over 7,000 cultivars): pomiferous.com

Purdue University College of Agriculture: ag.purdue.edu

Salt Spring Apple Company (very good descriptions of many apples): saltspringapplecompany.com

University of Maine School of Food and Agriculture: umaine.edu/foodandagriculture

University of Minnesota Extension: extension.umn.edu

University of New Hampshire Extension, "Fruit Injury Types Recognized in Annual New Hampshire Apple Harvest Evaluations" (excellent photos of insect injury): extension.unh.edu/resource/fruit-injury-types-recognized-annual-new-hampshire-apple-harvest-evaluations-fact-sheet

University of Vermont Extension Tree Fruit site: uvm.edu/extension/horticulture/tree-fruit

Vintage Virginia Apples and Albemarle Cider Works: albemarleciderworks.com

Index

Page numbers in italic represent photos.